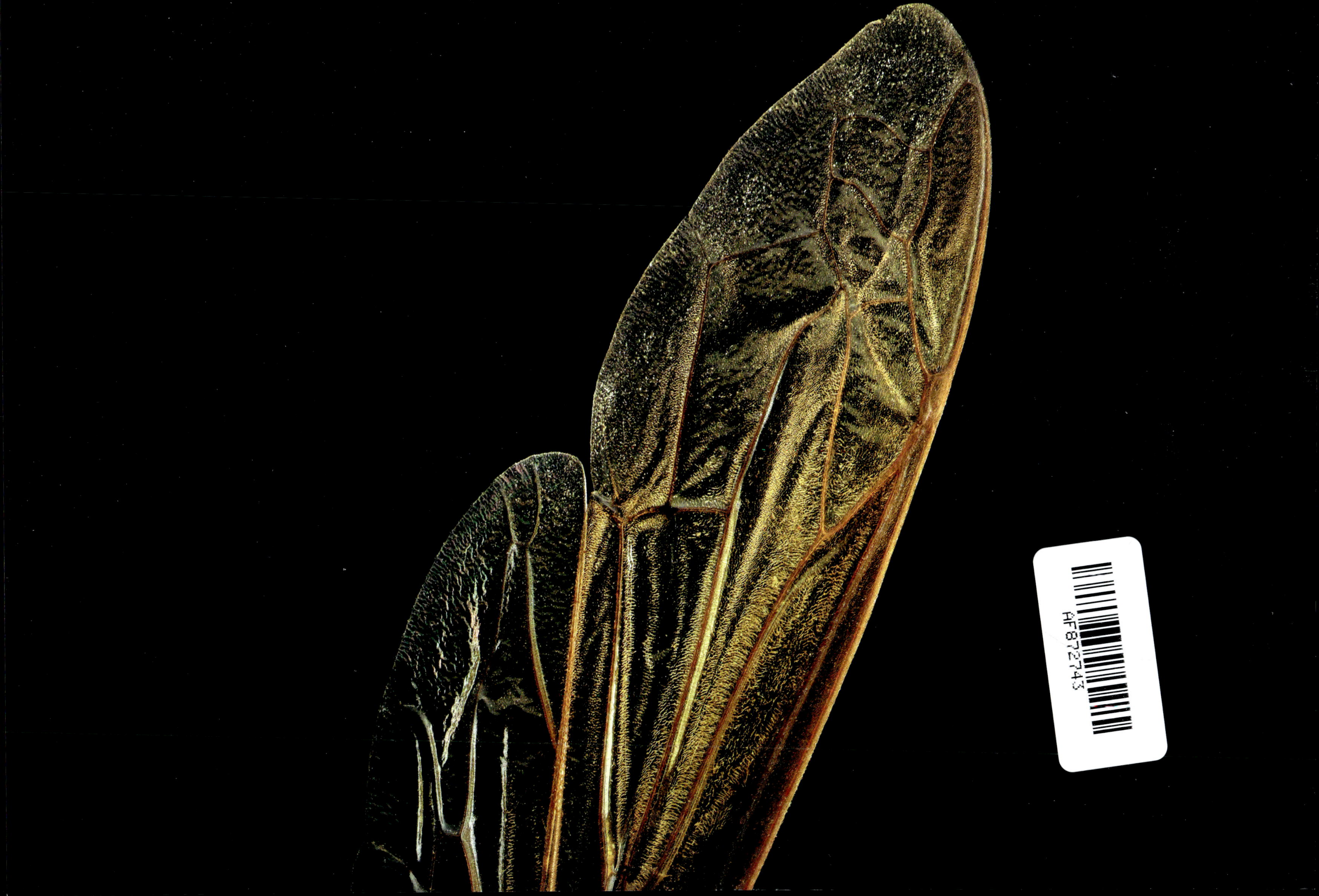
AF872743

microsculpture

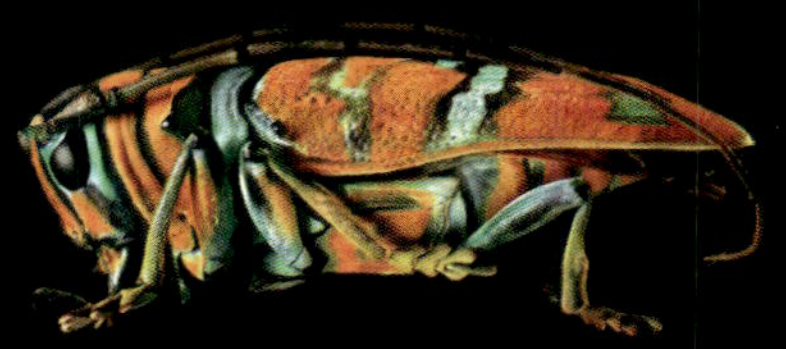

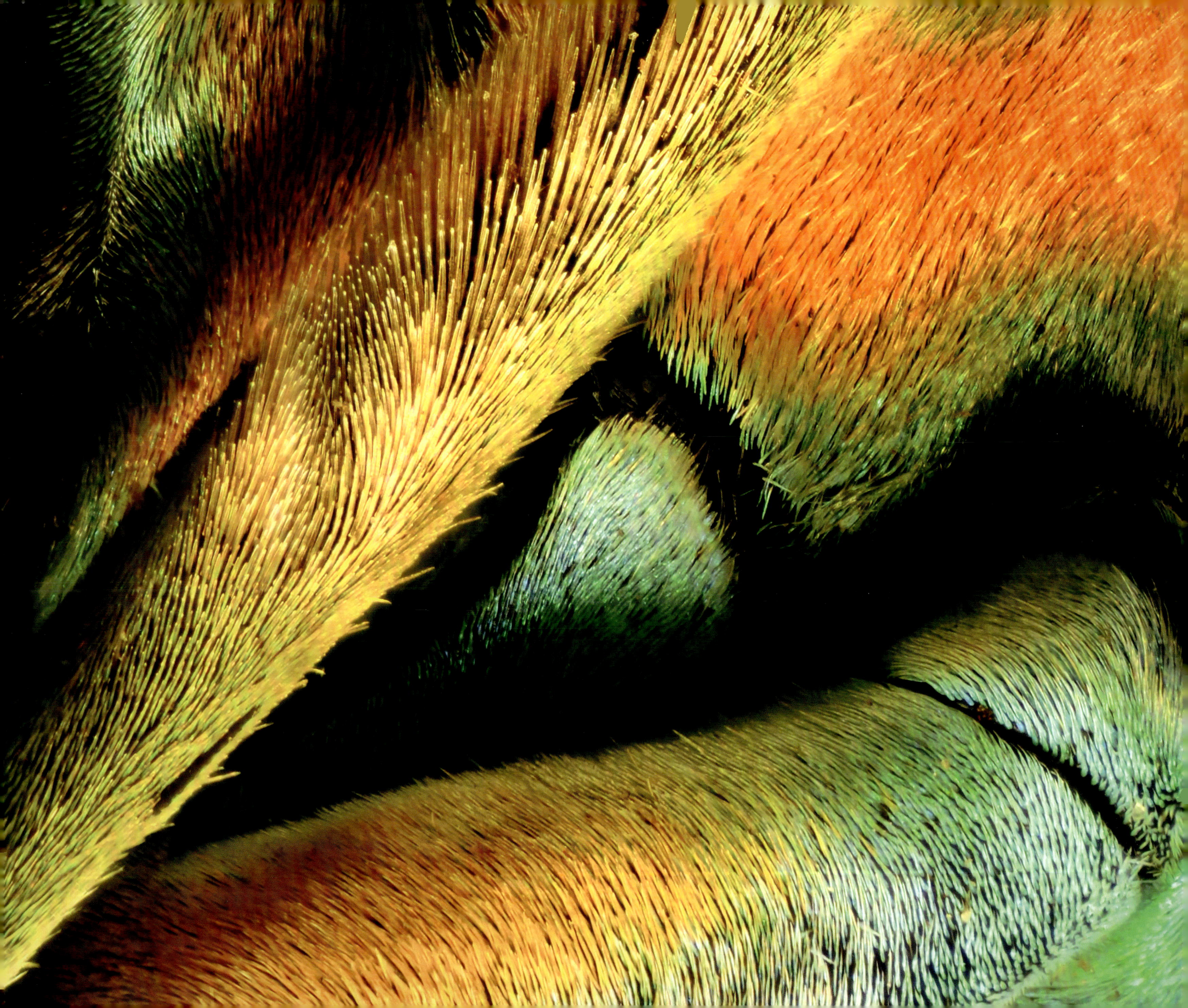

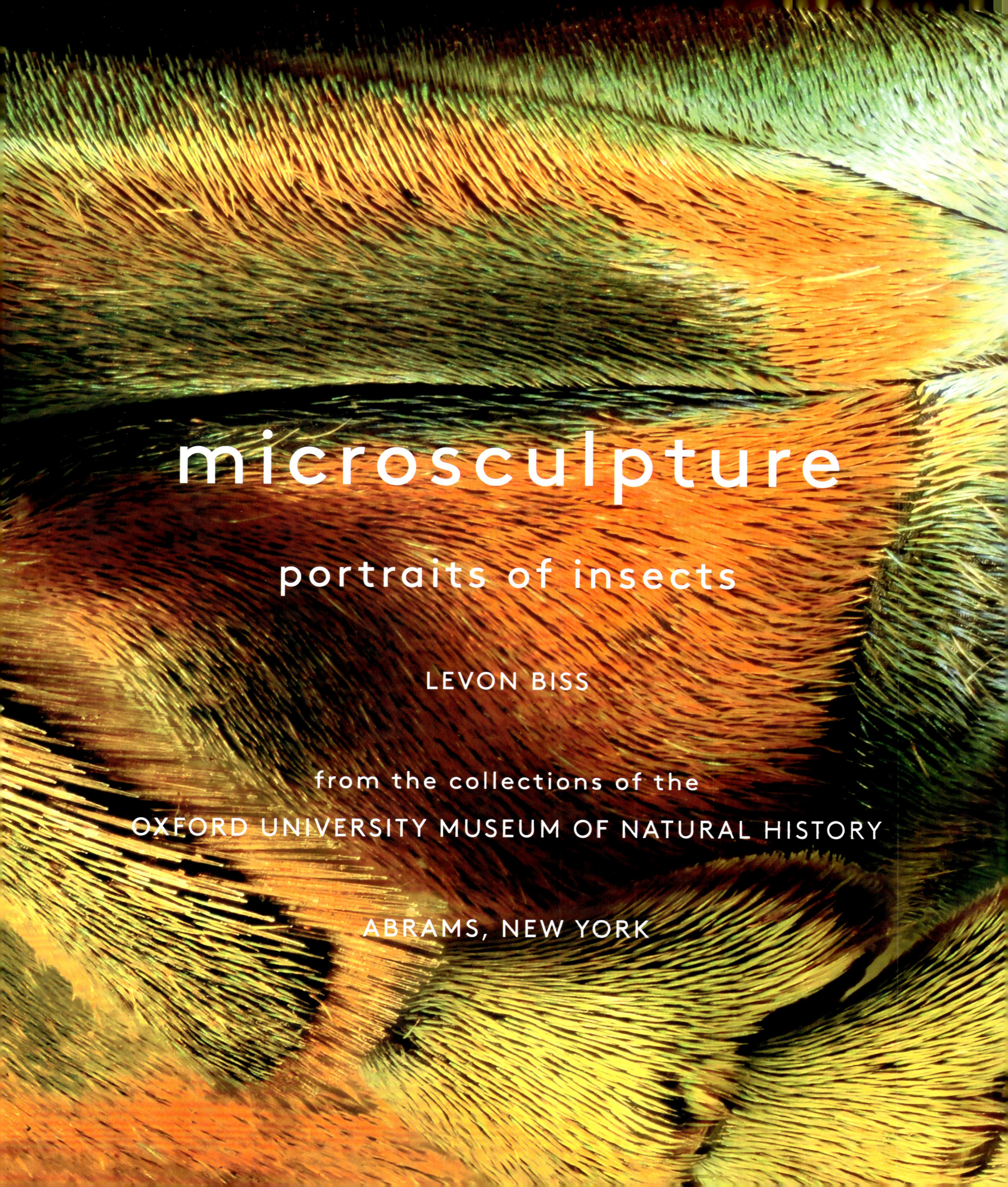

microsculpture

portraits of insects

LEVON BISS

from the collections of the

OXFORD UNIVERSITY MUSEUM OF NATURAL HISTORY

ABRAMS, NEW YORK

foreword

Insects are a vital part of our world. Most of their activities are obvious: Everyone is familiar with the busy hum of bees on a warm day or the tireless foraging of ants. Often, though, they go about their lives overlooked by us. For example, they help to recycle nutrients and, as part of food chains, insects provide food for many other larger animals. Without them, many ecosystems would be unable to function and human society would struggle to exist.

Insects exist at a perspective much smaller than our own. Beyond the resolution of our eyesight, the complex structure of life in this miniature world is hidden from us.

We take this knowledge for granted, but before the development of optical lenses and microscopes, the complexity of life at these small scales was unknown. When popularized for the first time in Robert Hooke's 1665 book, *Micrographia*, it caused a sensation. It presented a portfolio of everyday objects drawn at high magnification in unprecedented detail. Significantly, some of the most striking and iconic images from the book are of insects and their incredibly intricate anatomy.

The art of drawing has since become an essential tool in biology for describing and recording the natural world, and modern advances in digital imaging provide new and exciting possibilities. In *Microsculpture*, Levon Biss has developed his own photographic technique that succesfully captures the structural complexity of insects in a large format and at stunning resolution. Ultra-high detail images of insects are nothing new of course—images produced by the scanning electron microscope show details beyond even the most powerful optical lenses—but for all their detail, electron microscope images are often cold, devoid of color, and offer little in the way of aesthetic interest.

The insects in this book have been carefully chosen from among our collection of over five million at the Oxford University Museum of Natural History. Collections such as this represent a vast repository of information on biodiversity and are ever more valuable in the context of our rapidly changing world. Usually this collection is kept behind closed doors, most of the specimens being too small and fragile for public display. This book provides the perfect opportunity to showcase a few of its many and fascinating species.

The specimens we have chosen are from all over the world, from a backyard garden in England to a remote island near Antarctica. Some were collected more than one hundred and fifty years ago, some only recently—one was even collected by Charles Darwin himself during the voyage of the HMS *Beagle*. Together they proudly display the great diversity of insects: their unique shapes, dazzling colors, and all of their surface detail. To our eyes, the surfaces of insects appear smooth, but Levon's technique, as you will see in the pages of this book, reveals much more. At magnification, the insect's surface undergoes a transformation: Ridges, pits, and engraved meshes suddenly come into view, often overlaid with other minute structures such as hairs and scales.

These microstructures are thought to have many different functions. Surface features are thought to alter the physical properties of the insect's surface, helping to shed water, reduce friction, or reflect light. The fine structure of hairs can be modified to carry pollen, to grip surfaces as smooth as glass, or to detect the tiniest air movements. Butterflies, moths, beetles, and other insects may be covered with scales so small that they appear as dust to the naked eye. These scales may provide camouflage, insulation, and sometimes can reflect and scatter light to produce some of the most intense and vibrant colors seen in nature.

Surprisingly, in many cases the function of these tiny structures is not fully understood. The entire surface of most species of ground beetles found in this book are covered with a microscopic raised mesh-like pattern of microsculpture, giving the insects a matte or satin appearance. In related species, this pattern seems to be correlated with habitat; species from wet areas often have a transverse pattern, and those from saline habitats have a pattern of slightly raised bumps. It is believed that the microsculpture reduces friction or prevents water or dirt from sticking to the insect's surface. Even though we may not know all the answers, it is likely that most of these fascinating microstructures are adaptations, the result of the evolutionary process of natural selection. *Microsculpture* shows us that the study of these structures is not merely academic, but can have real-world applications in developing new and more efficient materials inspired by nature.

Despite all these theories of anatomy and function, this book represents much more than just a collection of scientific photographs. Levon's technique produces photographs of stunning depth and aesthetic beauty. His clever use of shadow creates mood and atmosphere: backlighting is used to accentuate particular structures, the light shining through thin areas of exoskeleton producing beautiful warm amber tones. It has been a joy to see Levon bring each new subject to life.

This collection of photographs is a celebration of nature and photographic art, but it also shows how science and art can be effectively combined to produce something new and exciting.

Dr. James Hogan
Entomology Department Curator
Oxford University Museum of Natural History

GROUND BEETLE

Carabus elysii (Coleoptera, Carabidae)

With iridescent colors and intricate microsculpture, *Carabus* ground beetles are some of the most attractive insects. Beyond the resolution of the most powerful optical lenses are further levels of complexity. The intense colors of this beetle are produced not by pigments but by the interaction between light and separate layers of chitin in the exoskeleton. These layers can even be preserved in fossils, the colors still present after millions of years. Since the beetle is nocturnal and is never usually exposed to sunlight, the function of the colors is still something of an enigma.

CHINA

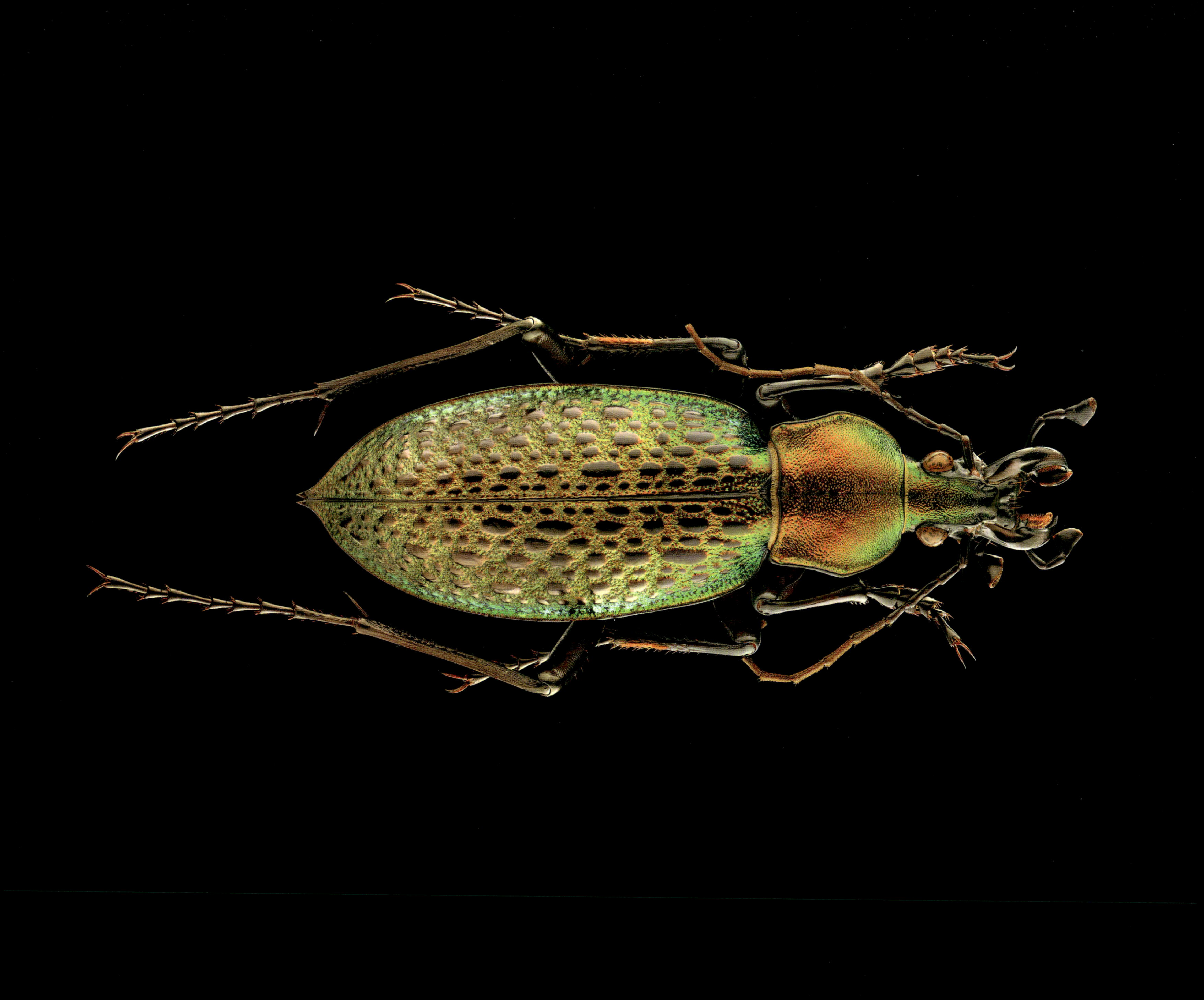

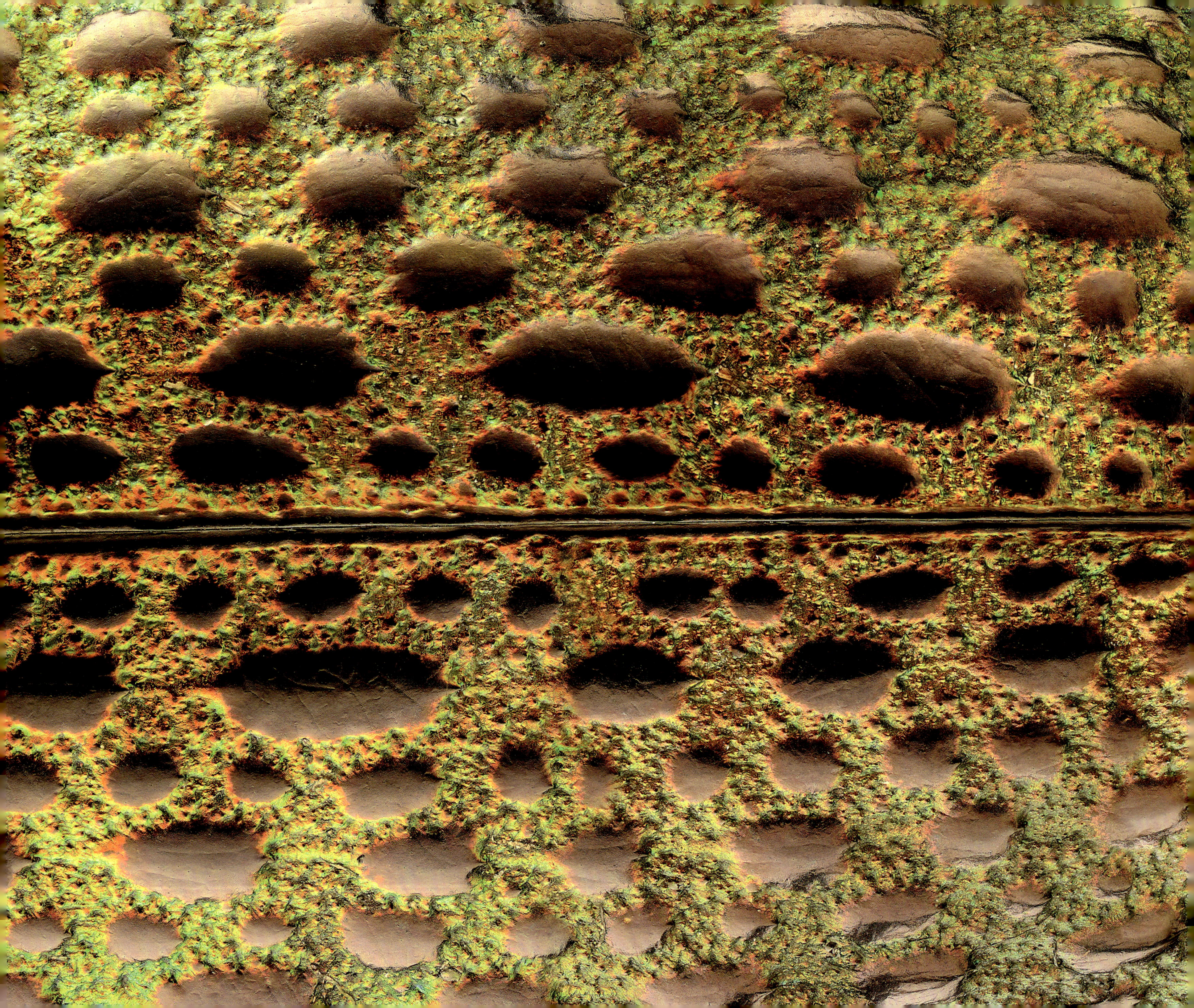

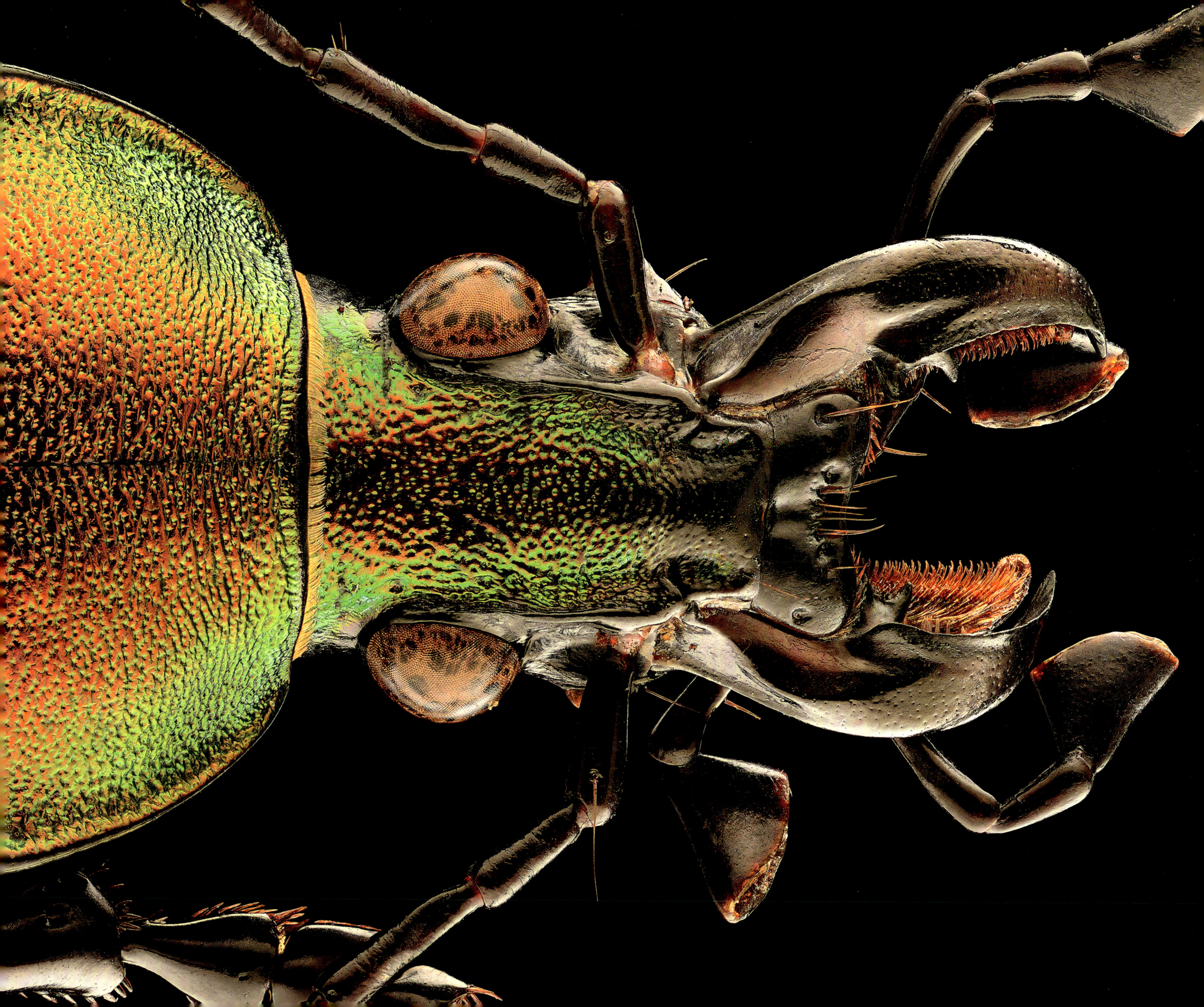

ASSASSIN BUG

SPECIES OF THE FAMILY Reduviidae (Hemiptera)

Most assassin bugs are efficient predators of other arthropods, particularly insects. As a group they are very diverse in form, the body sometimes armed with strange modifications such as toothed ridges or spikes. A few species have diversified into sucking mammalian blood and are responsible for spreading Chagas disease in humans. These are the so-called "kissing bugs," named for their habit of targeting the lips of their sleeping host.

BOLIVIA

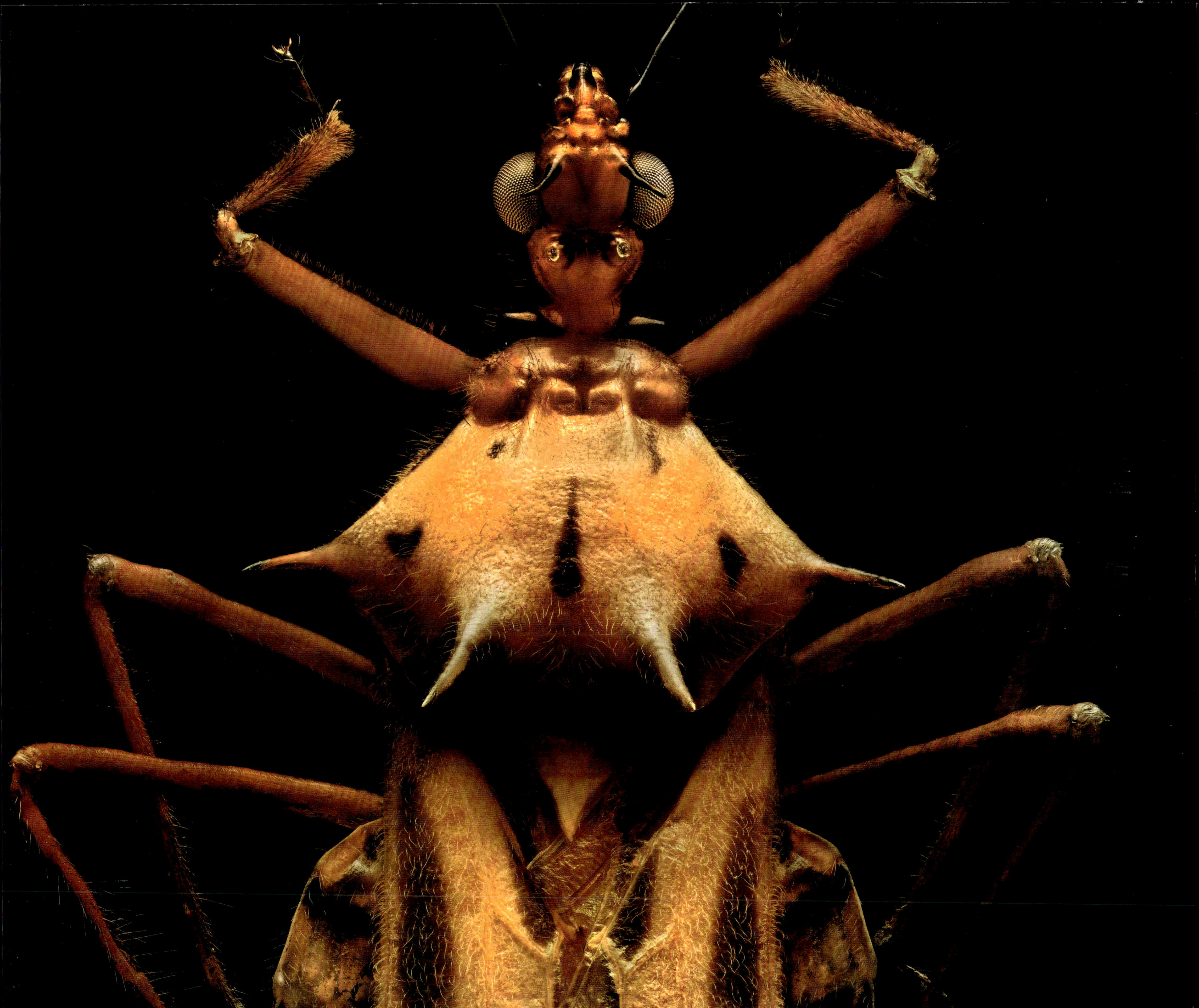

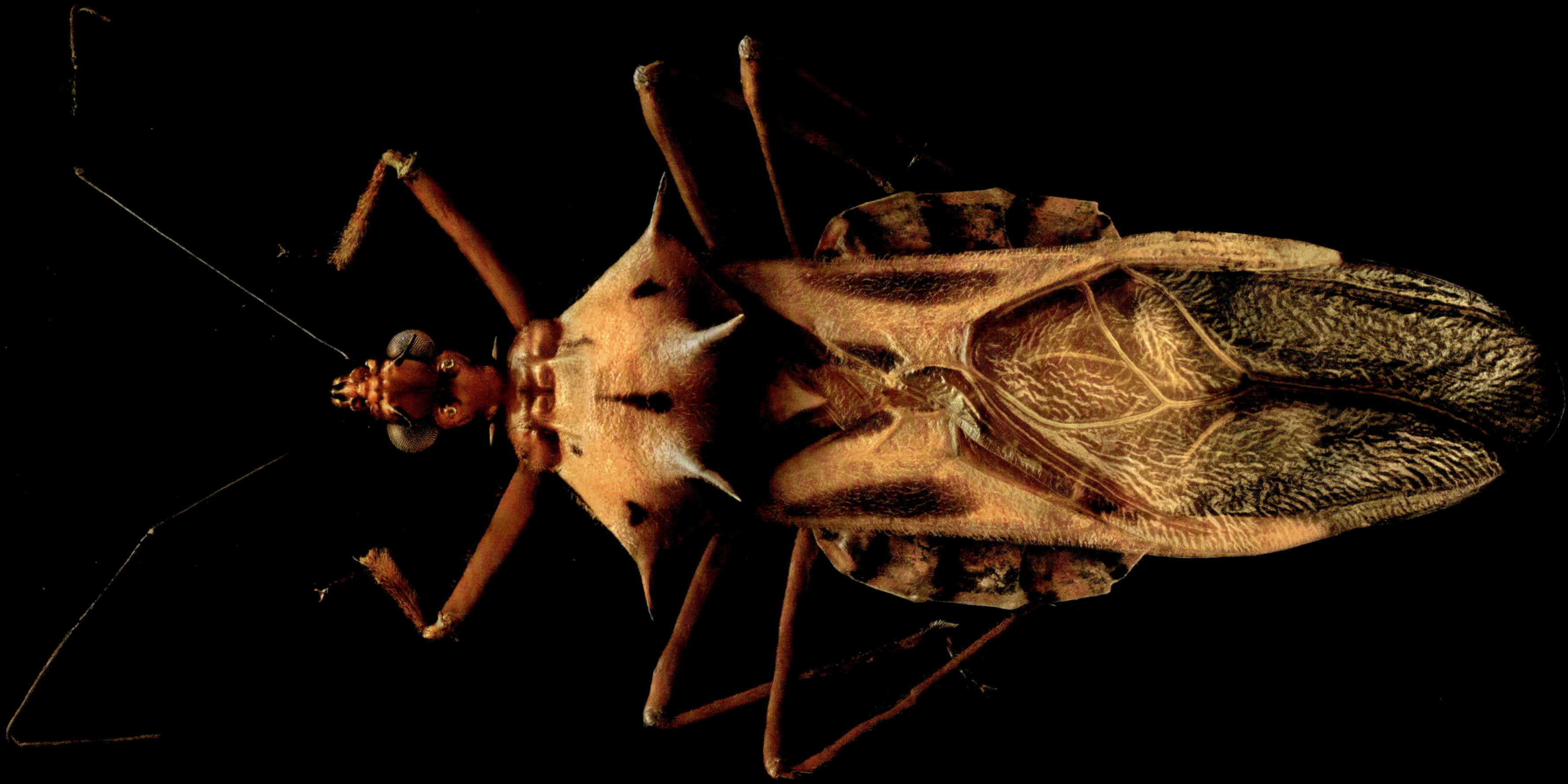

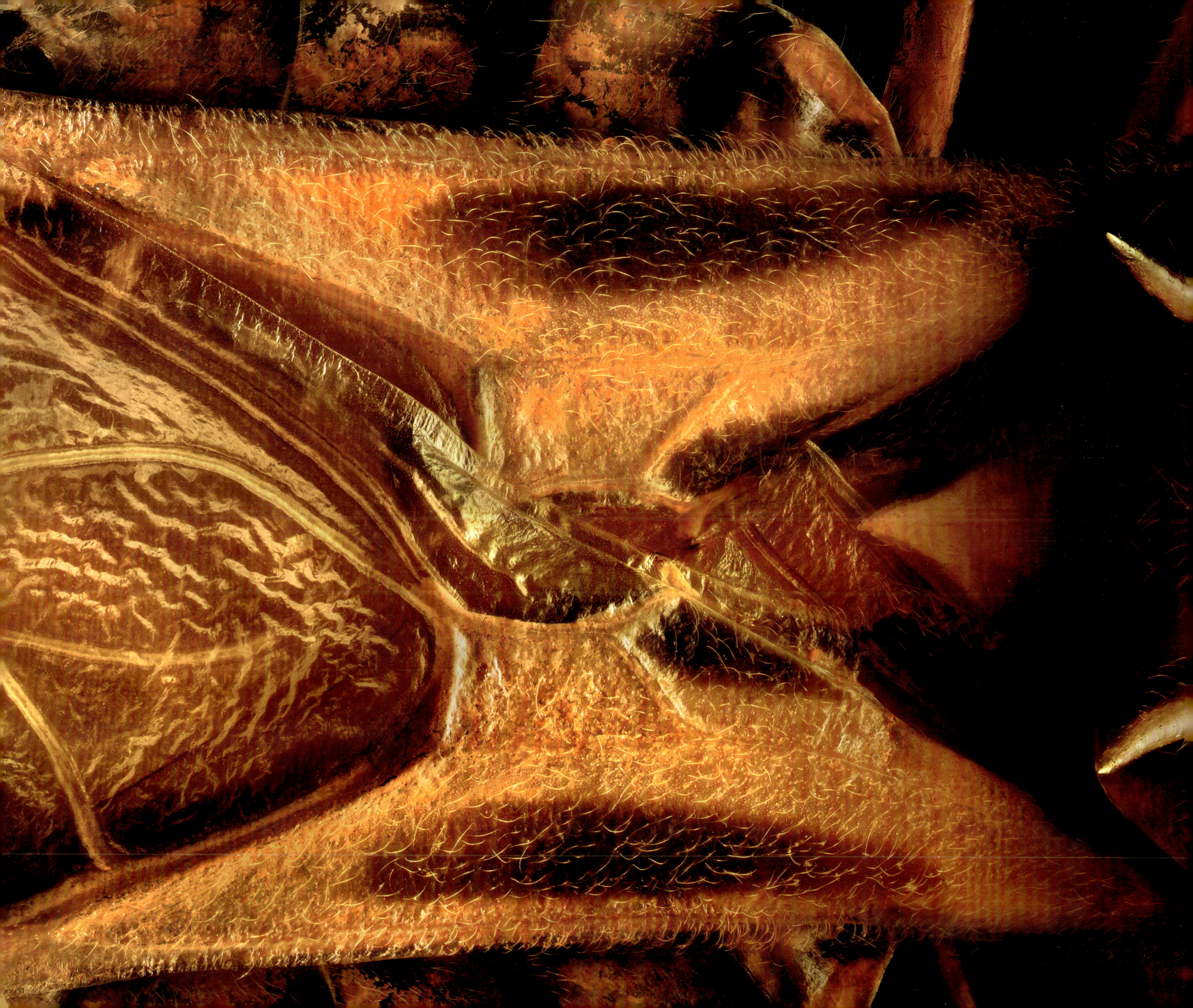

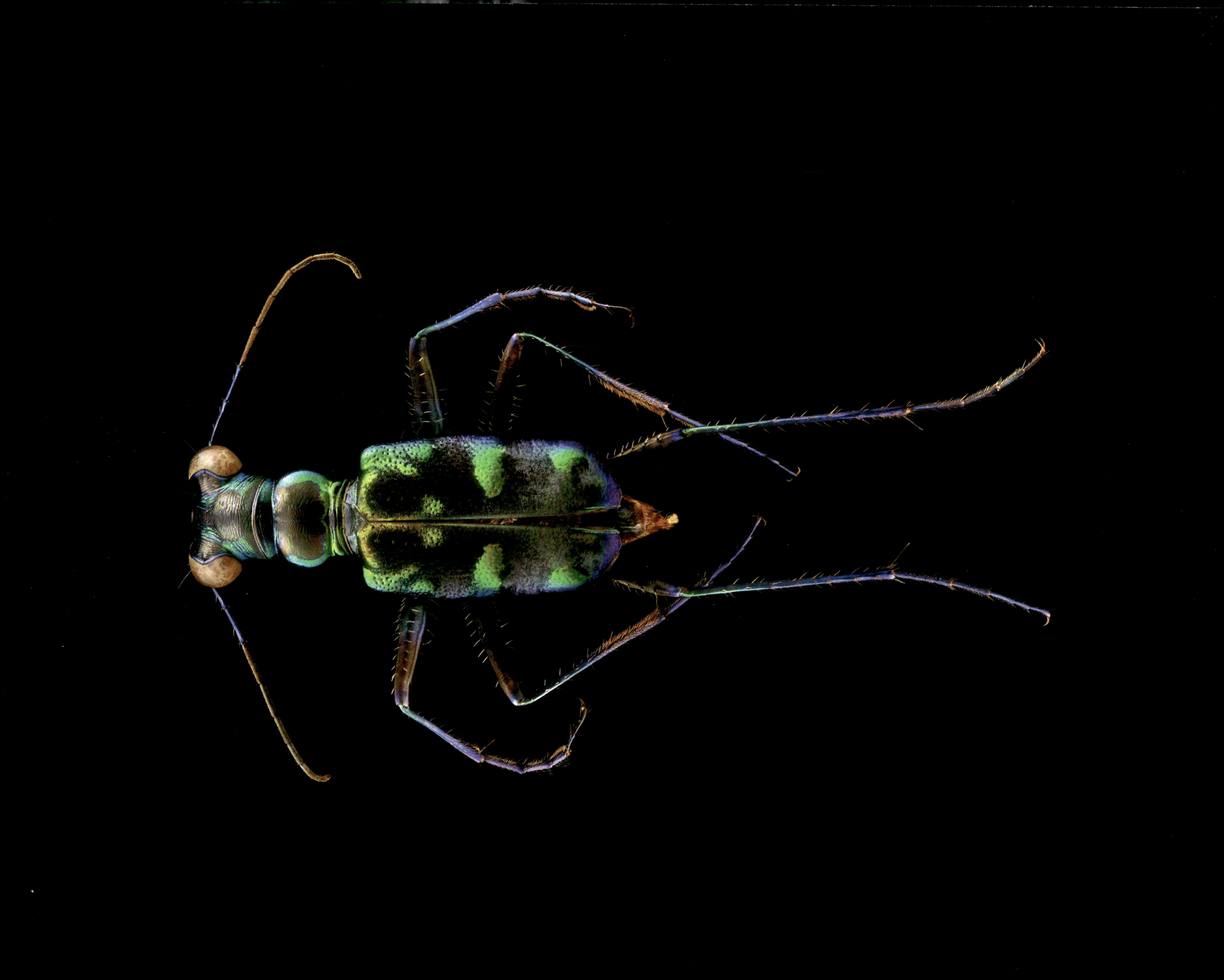

TIGER BEETLE

SPECIES OF THE SUBFAMILY Cicindelinae
(Coleoptera, Carabidae)

Tiger beetles are ferocious and powerful hunters in miniature, using their large bulging eyes and long legs to detect and run down their prey. Viewed closely, the surface of the insect has a complex microsculpture of minute pits overlaid with an even smaller mesh-like surface. Underneath these structural elements appears an almost three-dimensional color pattern in varying shades of green, probably to provide camouflage in its rainforest habitat.

BORNEO

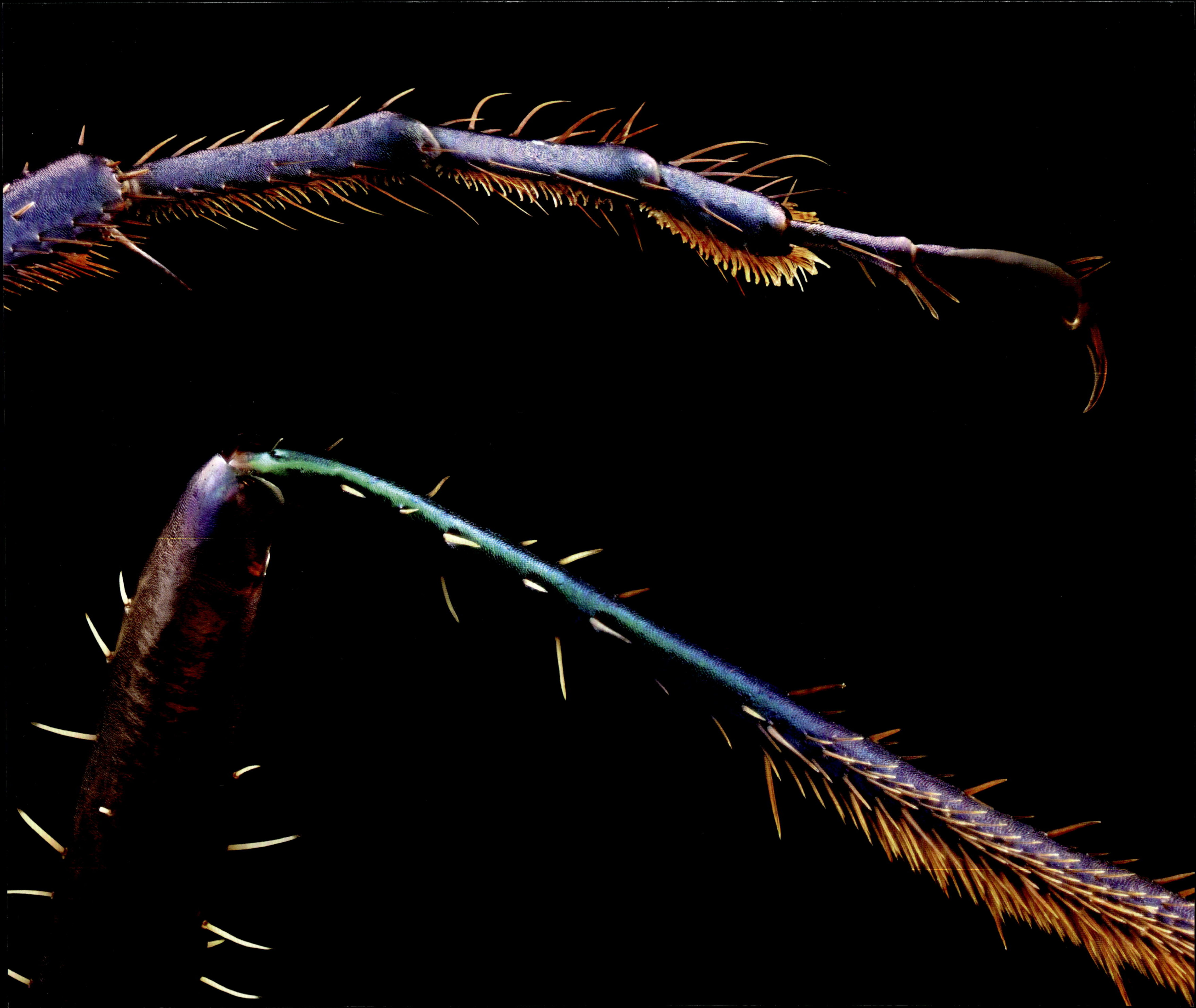

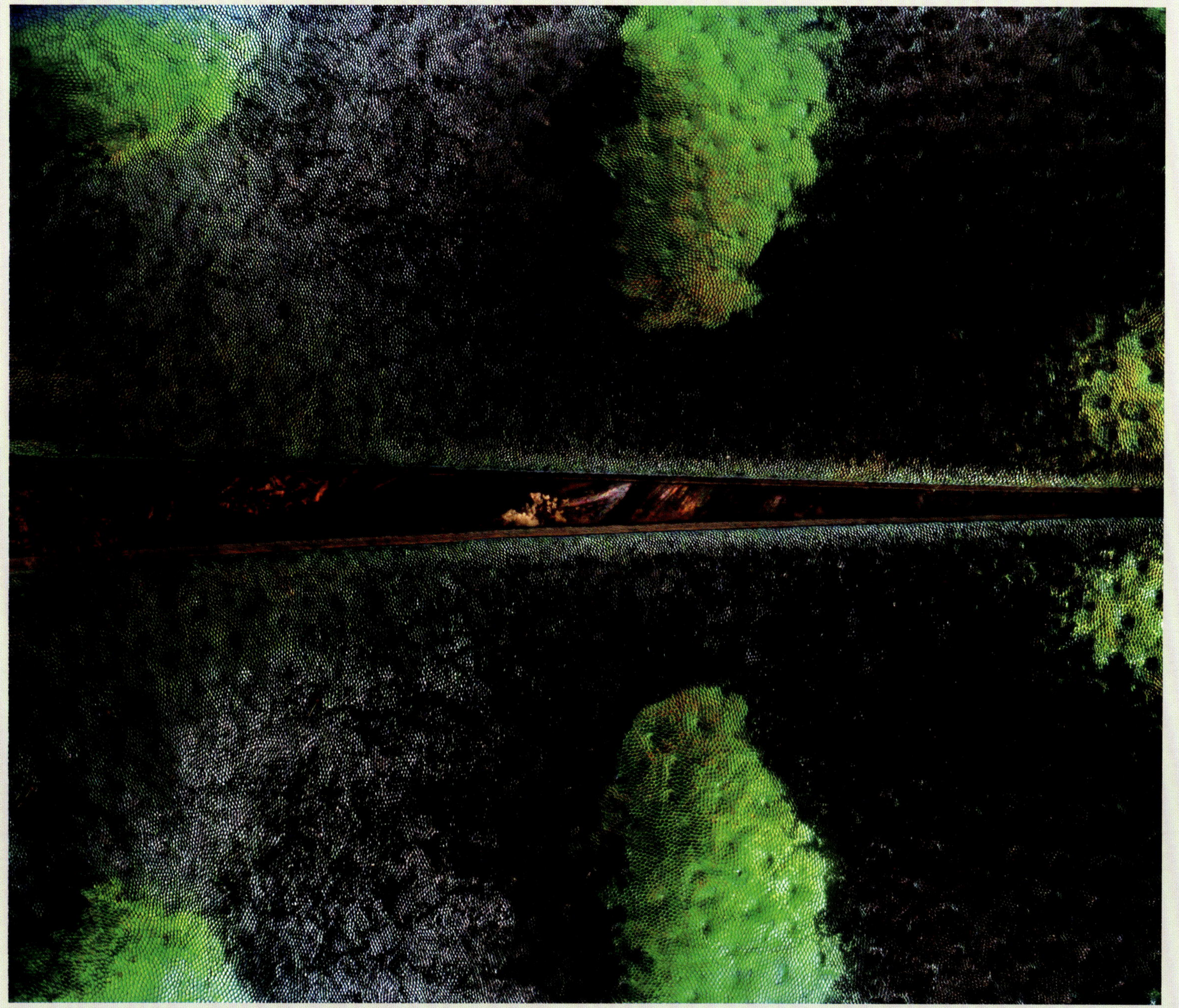

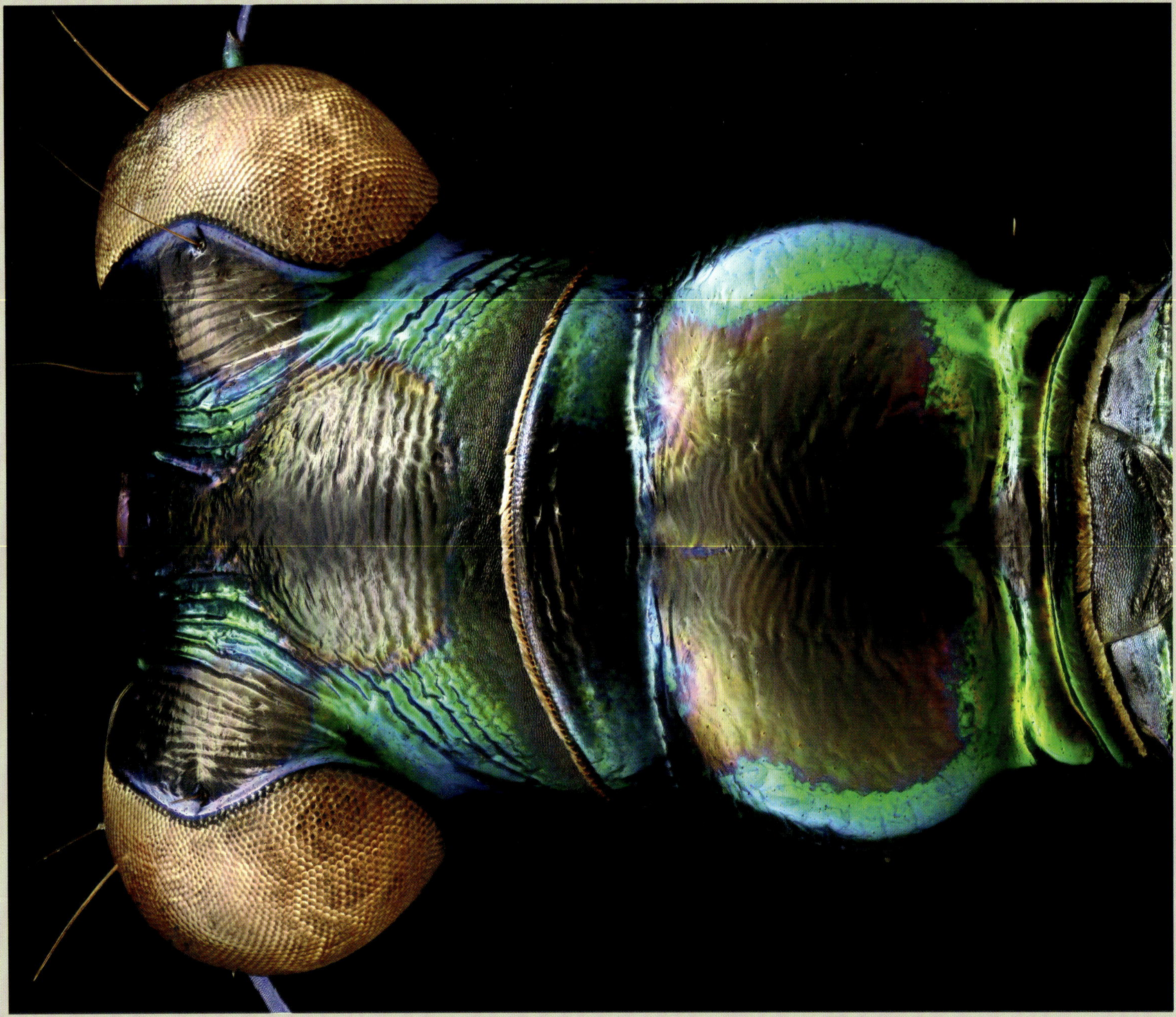

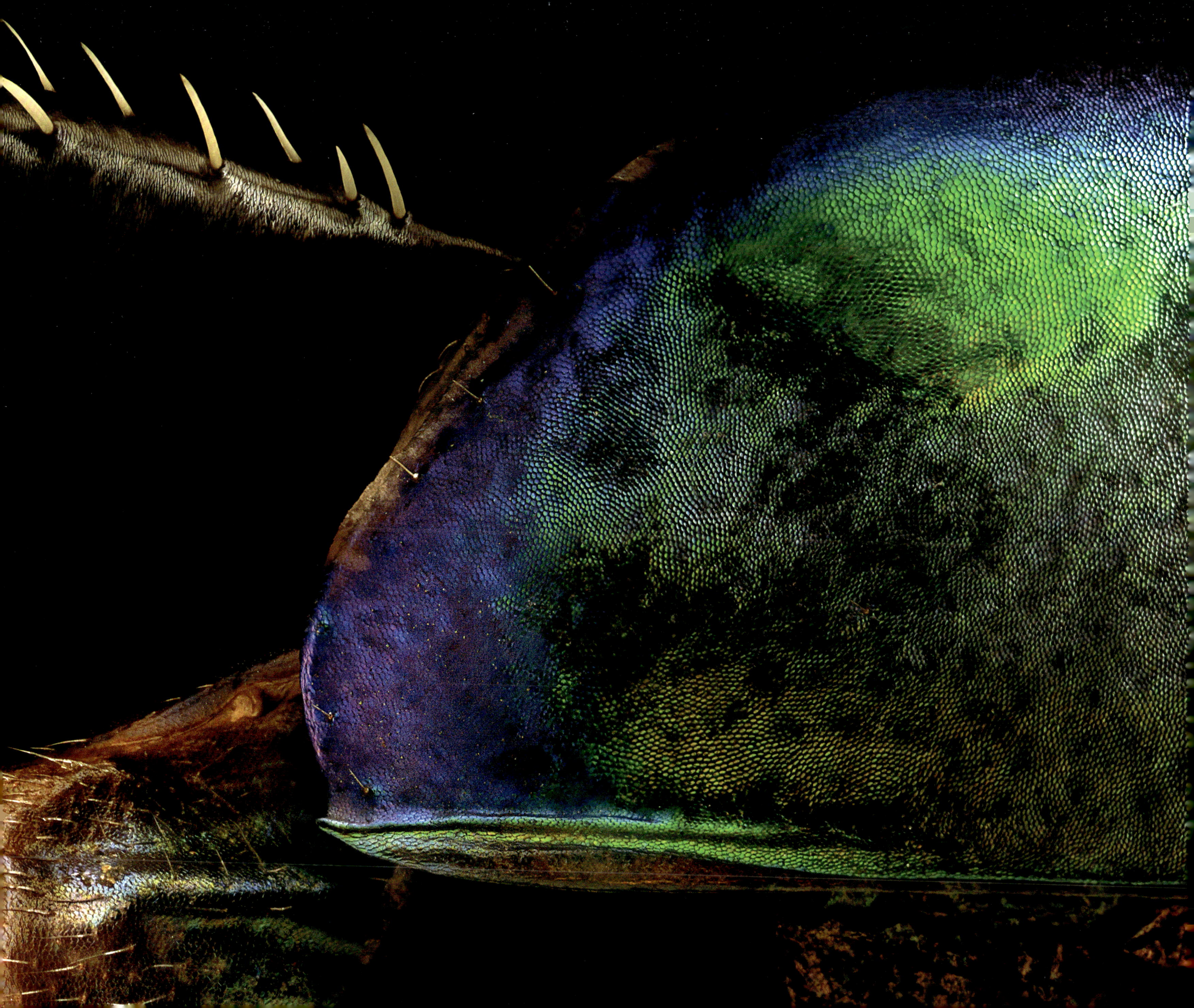

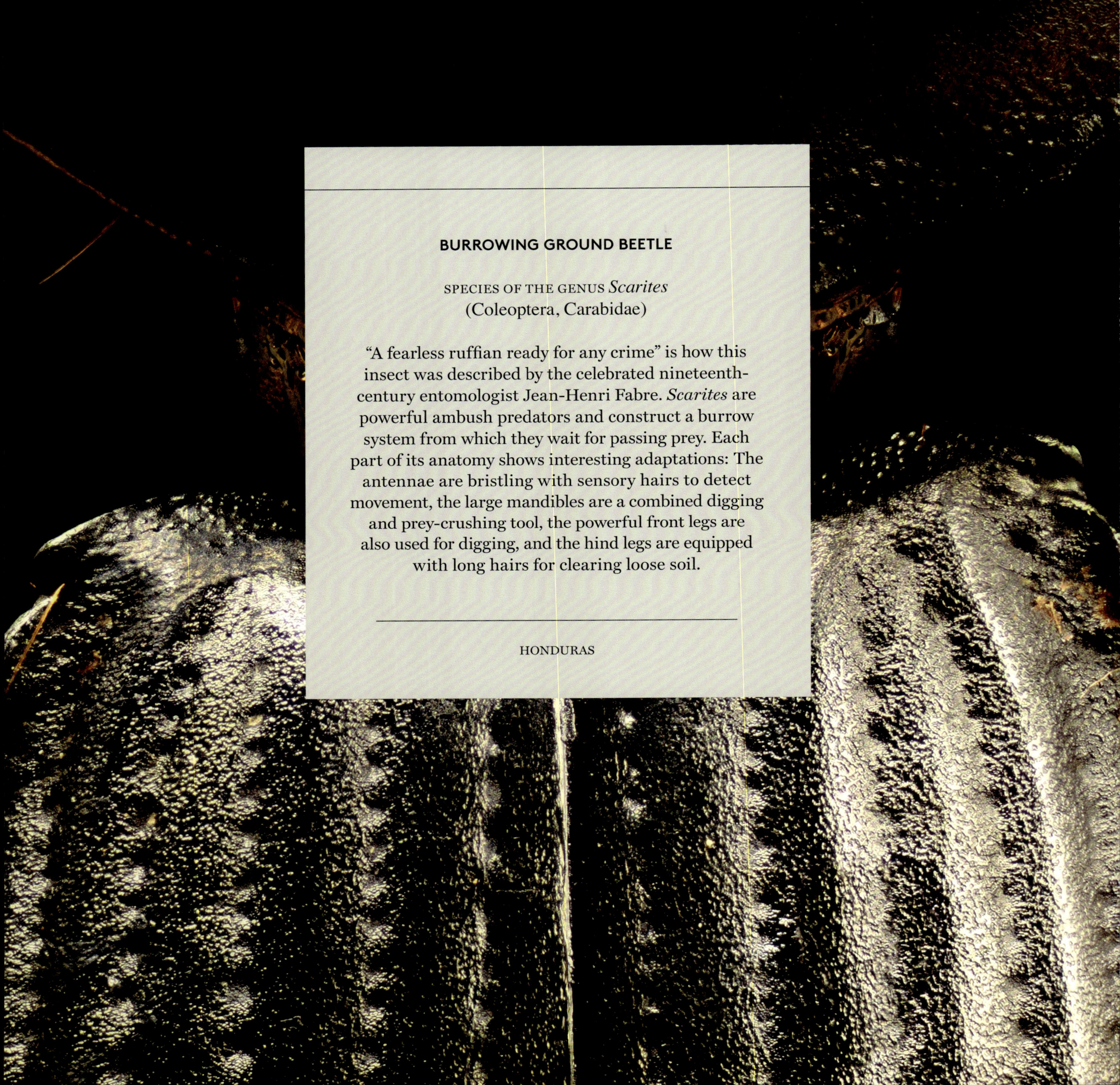

BURROWING GROUND BEETLE

SPECIES OF THE GENUS *Scarites*
(Coleoptera, Carabidae)

"A fearless ruffian ready for any crime" is how this insect was described by the celebrated nineteenth-century entomologist Jean-Henri Fabre. *Scarites* are powerful ambush predators and construct a burrow system from which they wait for passing prey. Each part of its anatomy shows interesting adaptations: The antennae are bristling with sensory hairs to detect movement, the large mandibles are a combined digging and prey-crushing tool, the powerful front legs are also used for digging, and the hind legs are equipped with long hairs for clearing loose soil.

HONDURAS

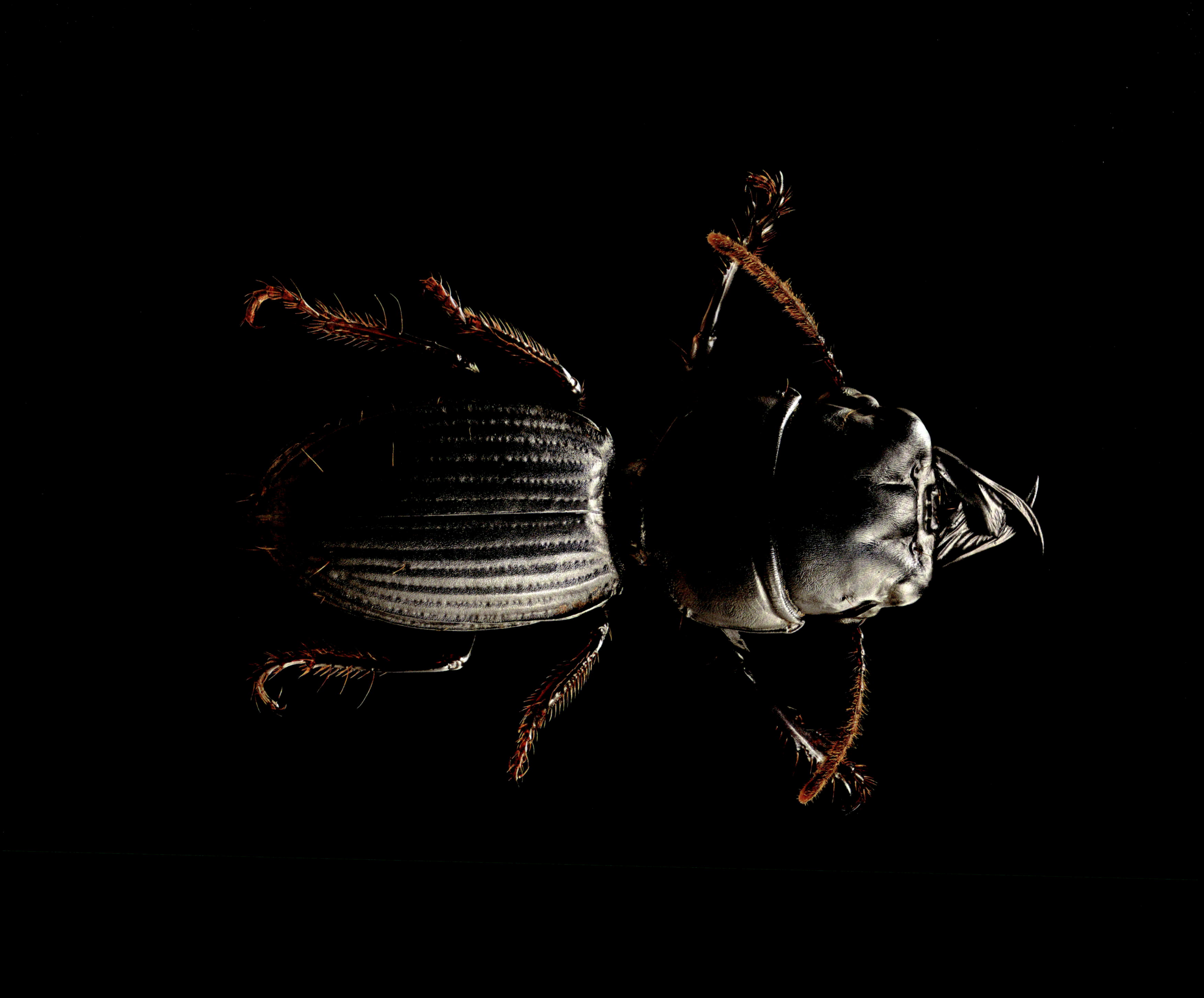

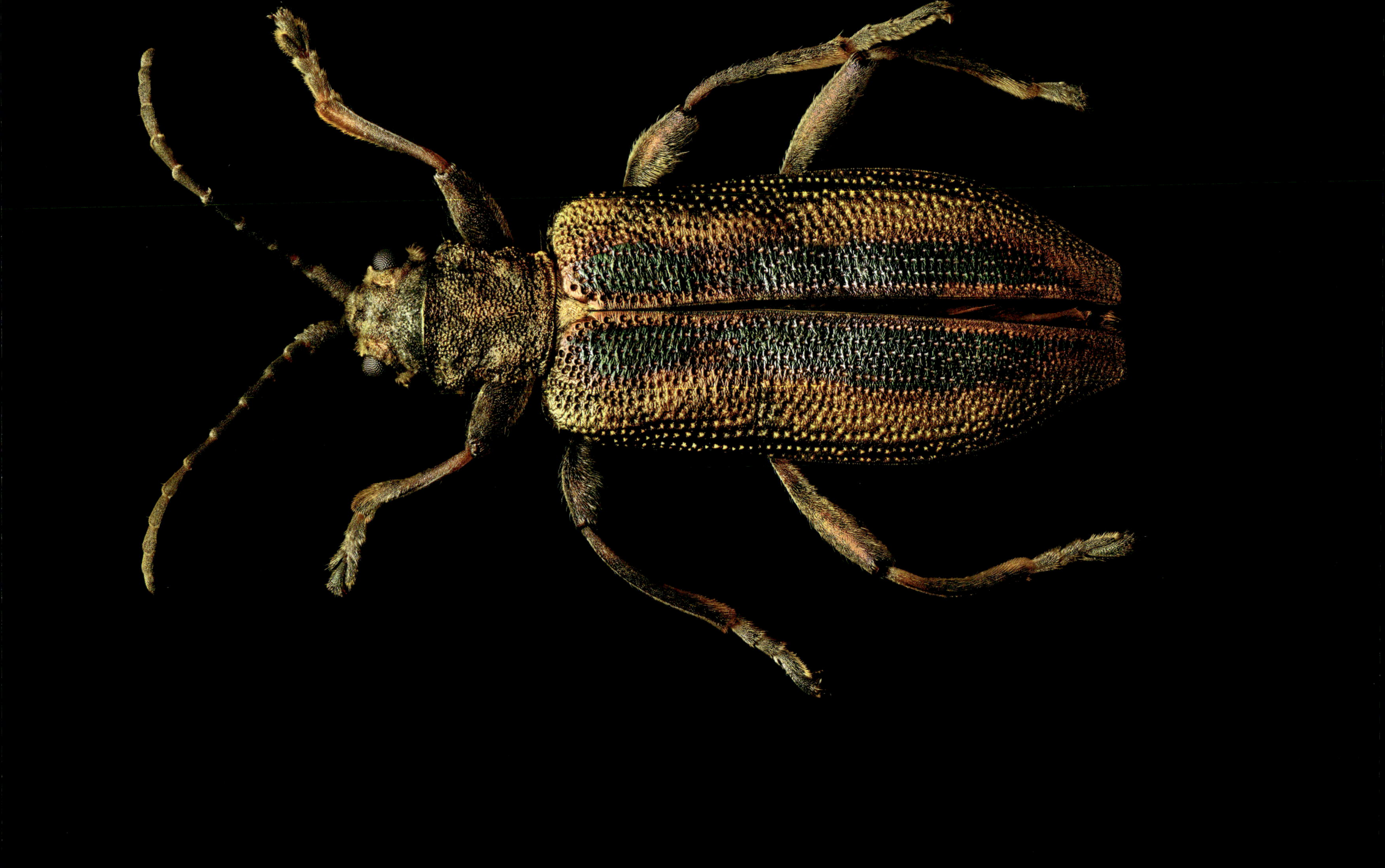

COMMON REED BEETLE

Donacia vulgaris (Coleoptera, Chrysomelidae)

This is a beetle of British wetlands with beautiful but subtle metallic stripes of color. The larvae of reed beetles have a highly unusual lifestyle, developing entirely under-water attached to the roots of aquatic plants. In order to breathe they tap into air spaces inside the plant using two hollow spikes at the end of the abdomen. When mature they pupate inside a cocoon, spun using a viscous fluid produced by specialized symbiotic bacteria living in the gut. The microsculpture of the upper surface of reed beetles is usually well developed. In this species it is manifest as a thick mesh-like pattern, giving the beetle a slightly dulled appearance reminiscent of tarnished metal.

ENGLAND

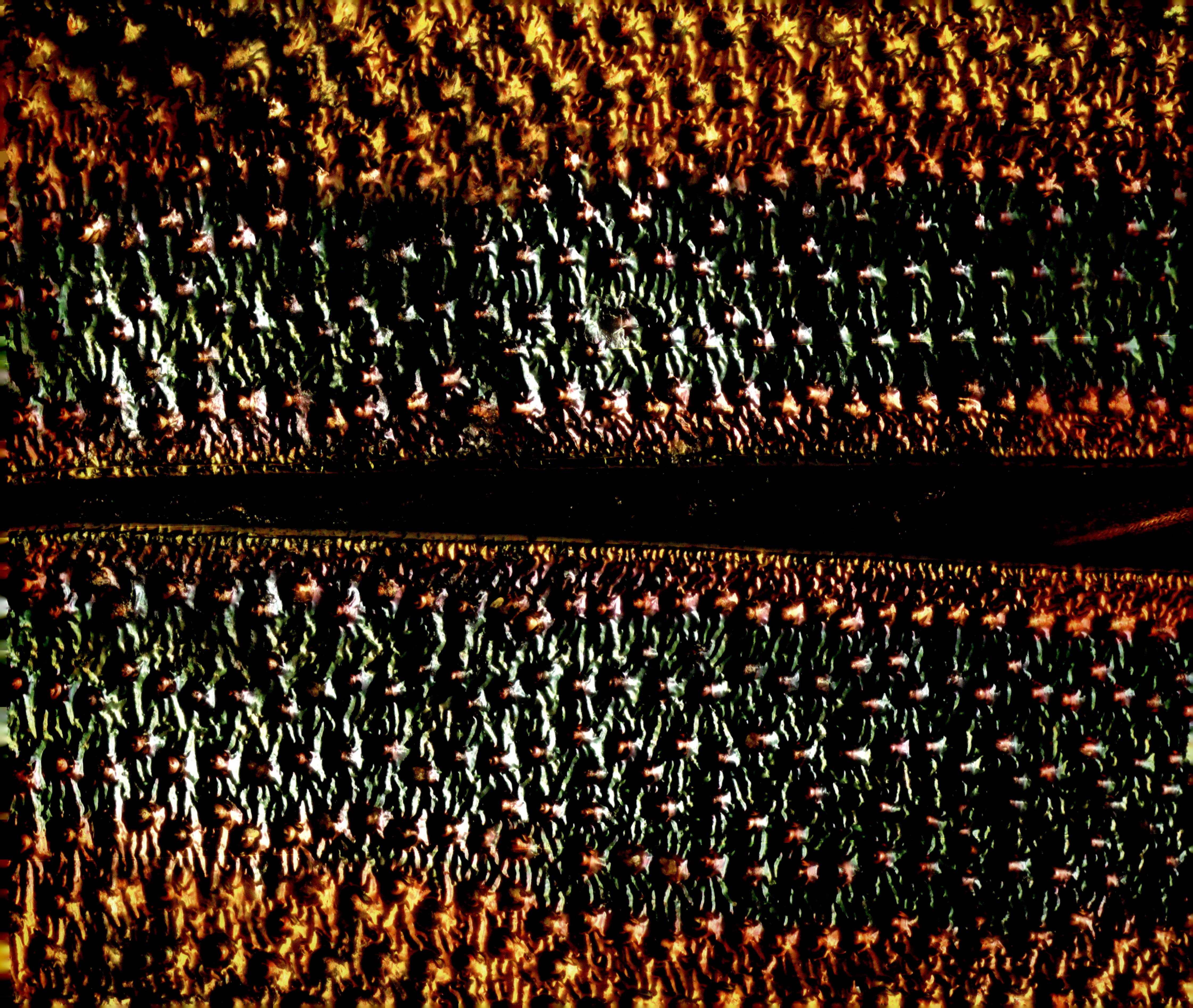

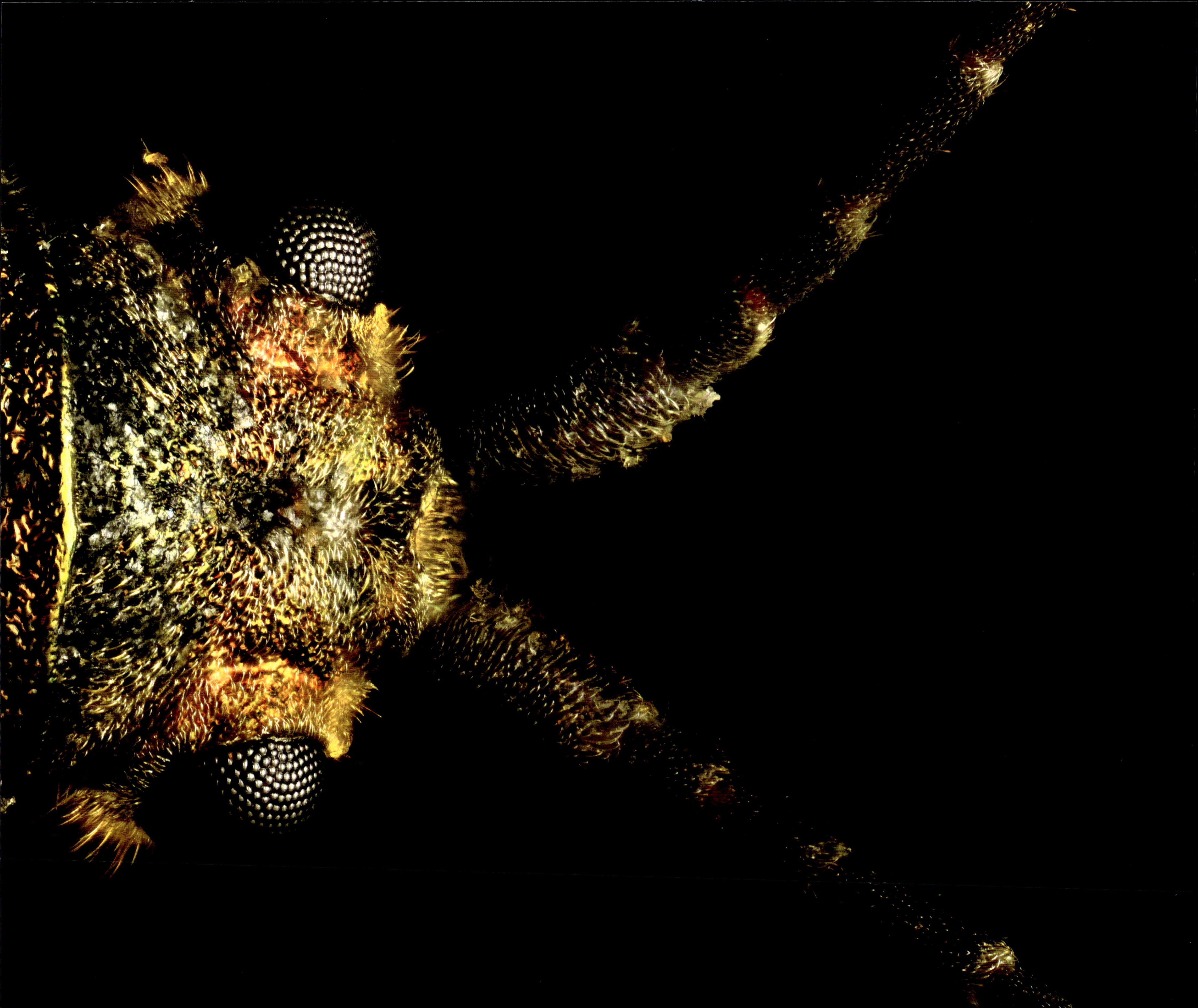

ORANGE NET-WINGED BEETLE

Lycus constrictus (Coleoptera, Lycidae)

The camera reveals the complex surface of the wing cases: long thickened ridges, a crossways net-like pattern, and two spikes, all overlain by a layer of extremely fine hairs. Lycid beetles produce potent defensive chemicals to deter attacks by predators, advertising this fact with distinctive orange and black warning colors. A wide range of harmless insects, including other beetles, wasps, flies, and moths, mimic the shape and color of lycids. Effectively employing false advertising, these mimics gain protection for free.

ZIMBABWE

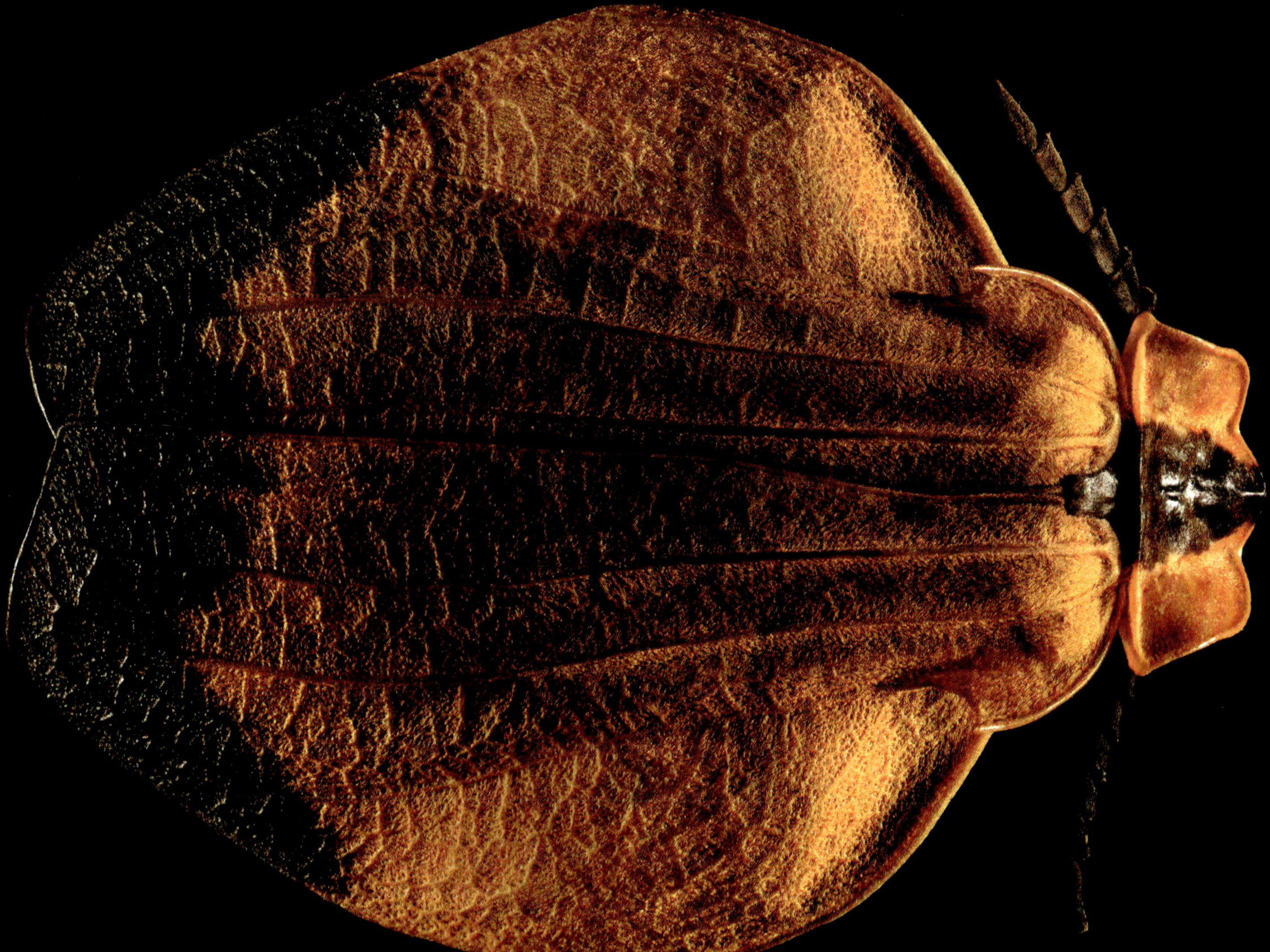

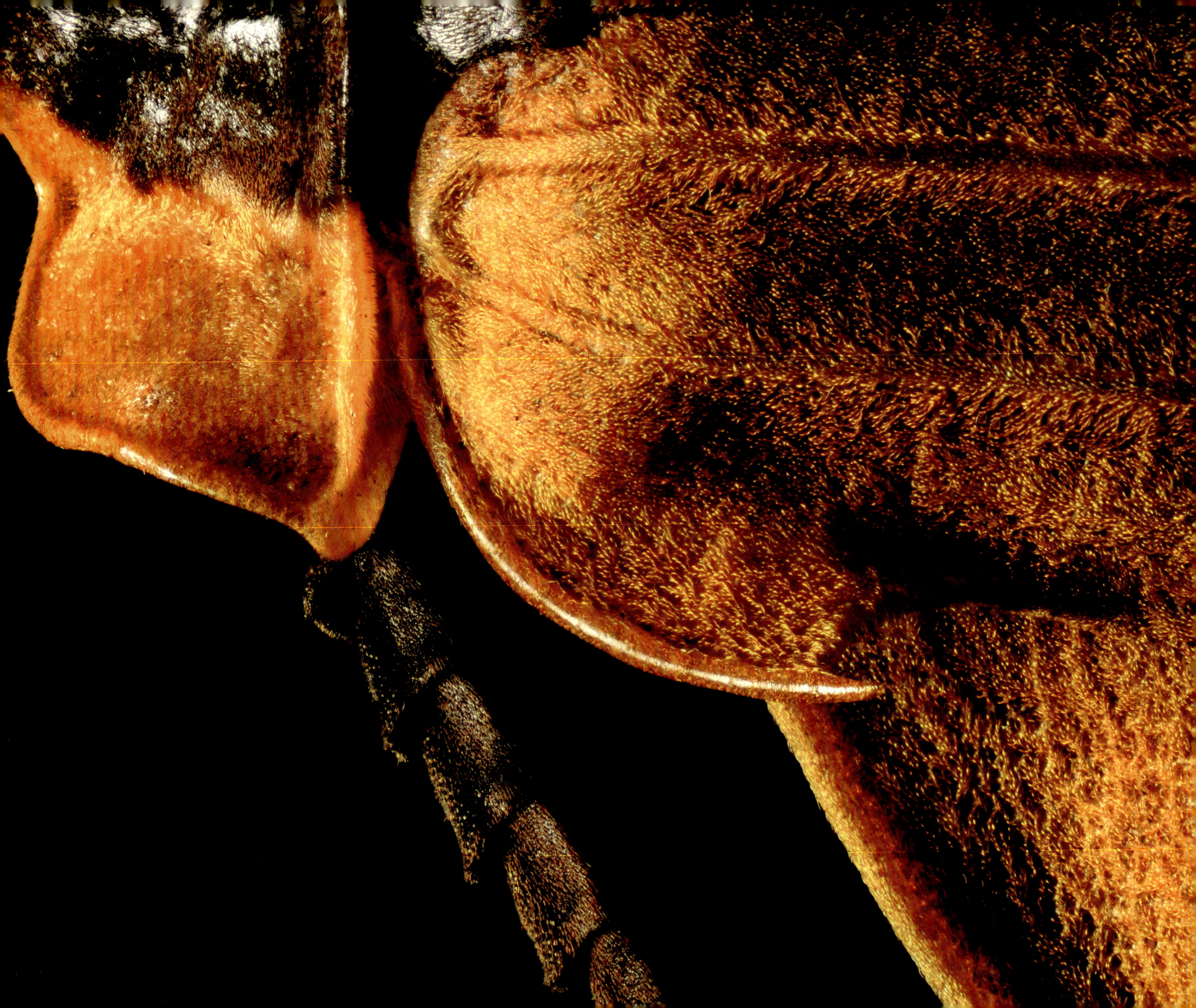

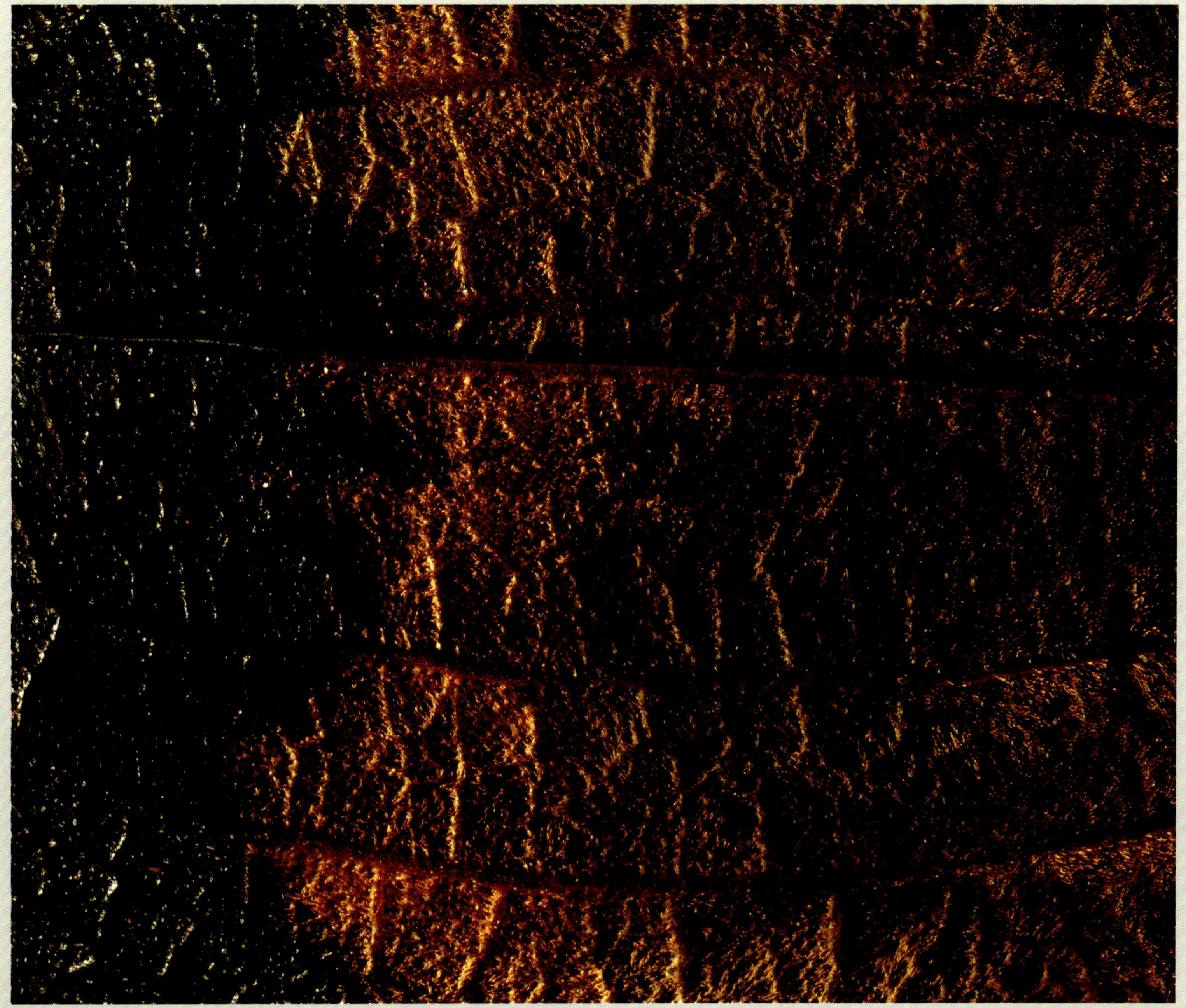

BRANCH-BACKED TREEHOPPER

SPECIES OF THE GENUS *Cladonota*
(Hemiptera, Membracidae)

A bizarre extension of the thorax gives this treehopper a strange appearance and probably helps to camouflage the insect. Once it was thought that this (and similar structures in other treehoppers) were simple outgrowths of the cuticle, but recent work on the development of treehoppers revealed something completely unexpected. The unusual outgrowths are in fact a modified *third* pair of wings, something exceptional and not seen in any other group of insects.

BELIZE

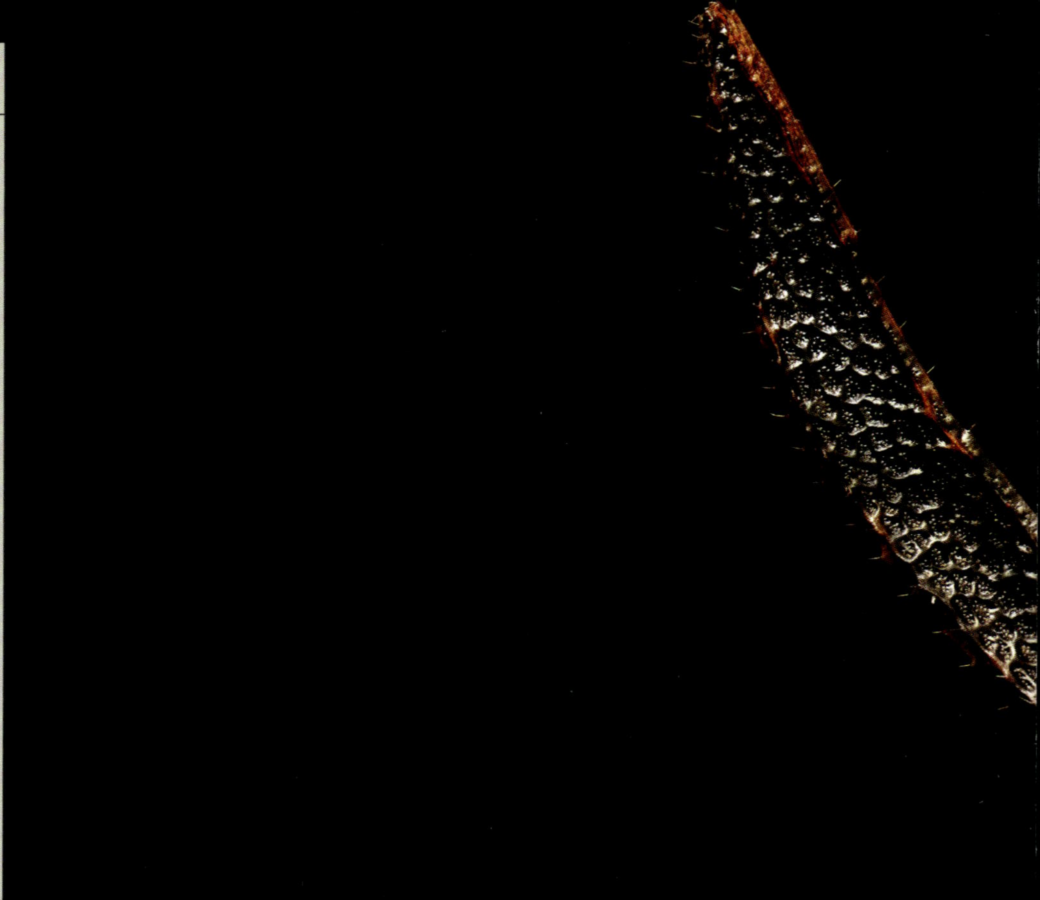

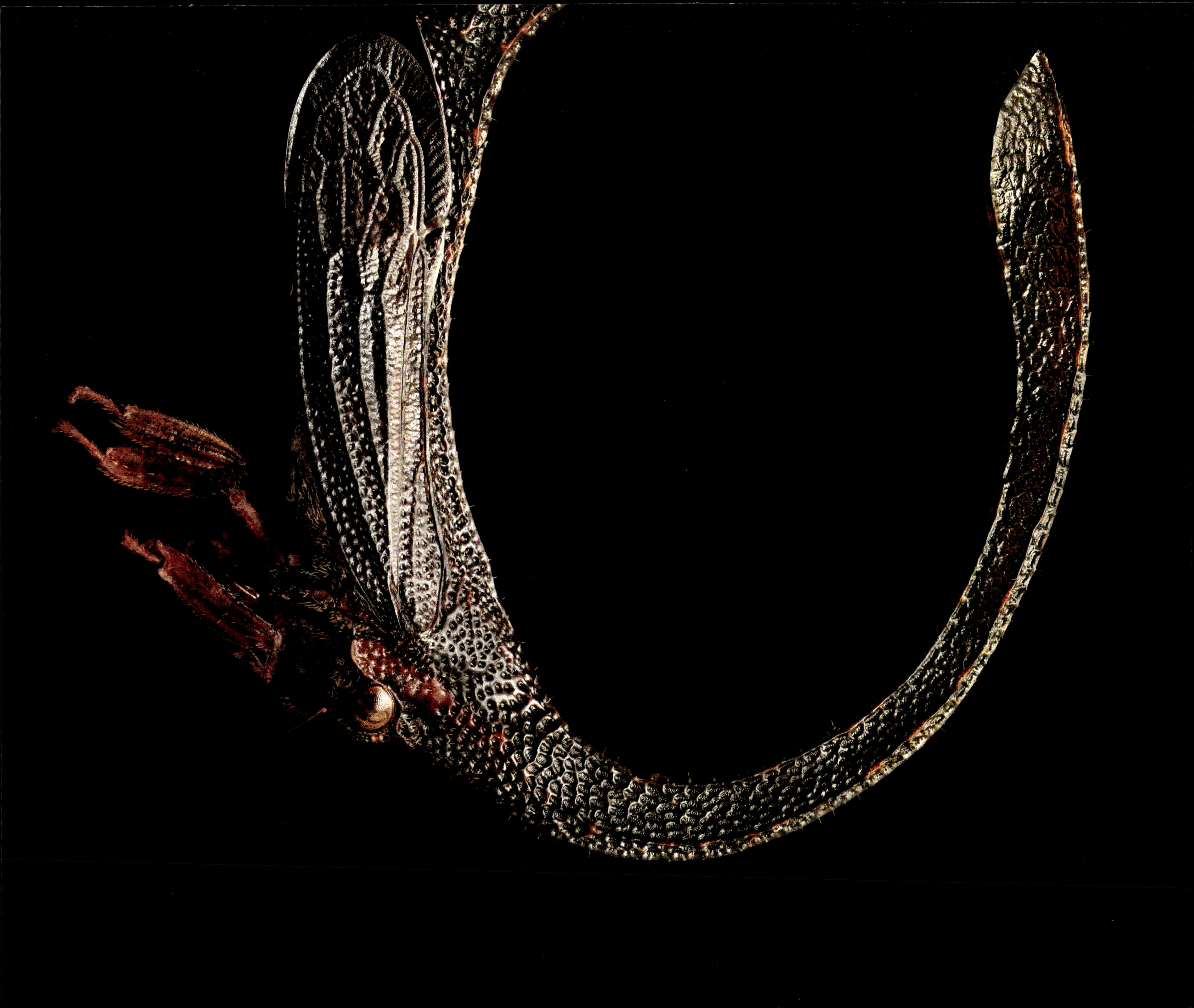

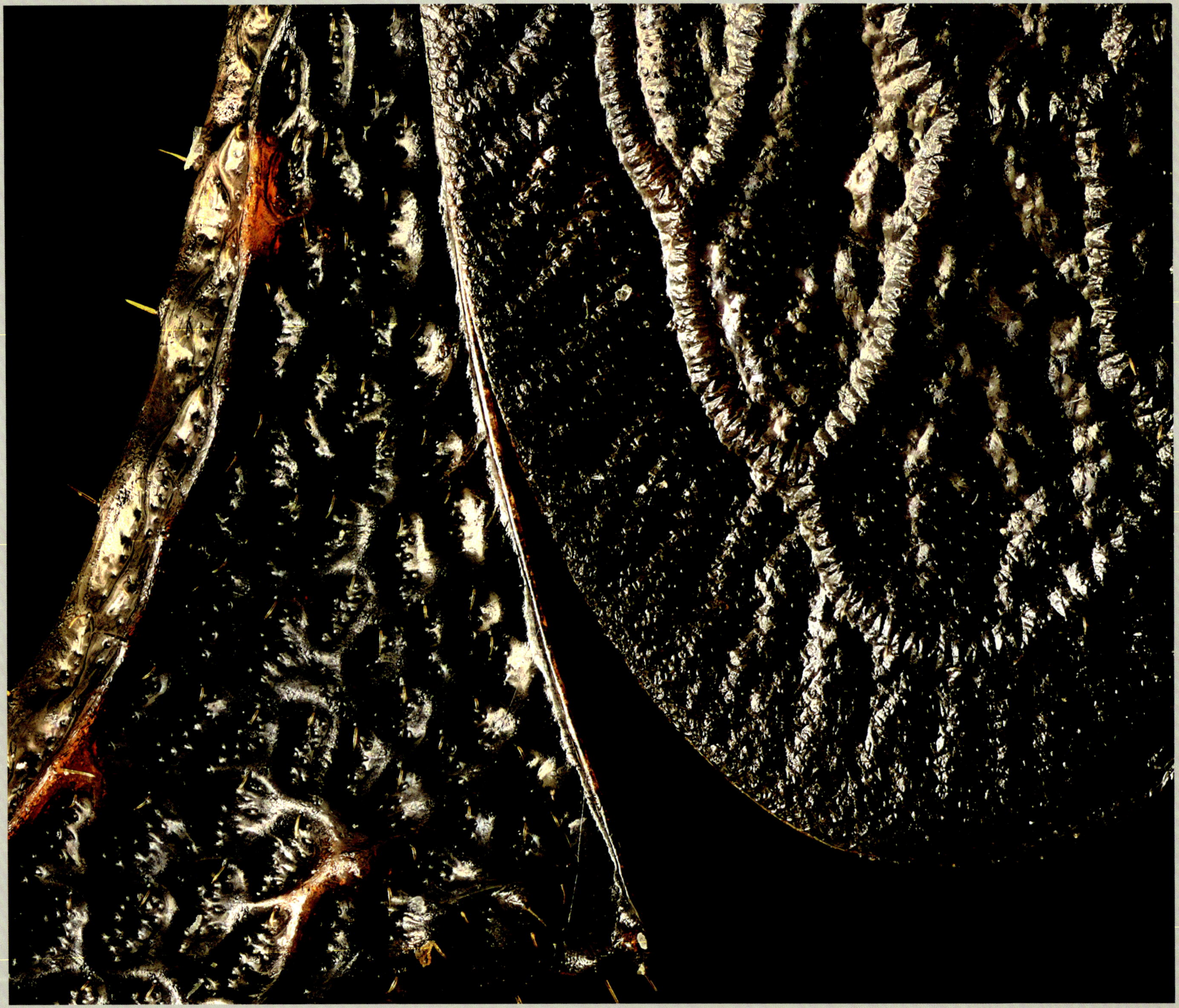

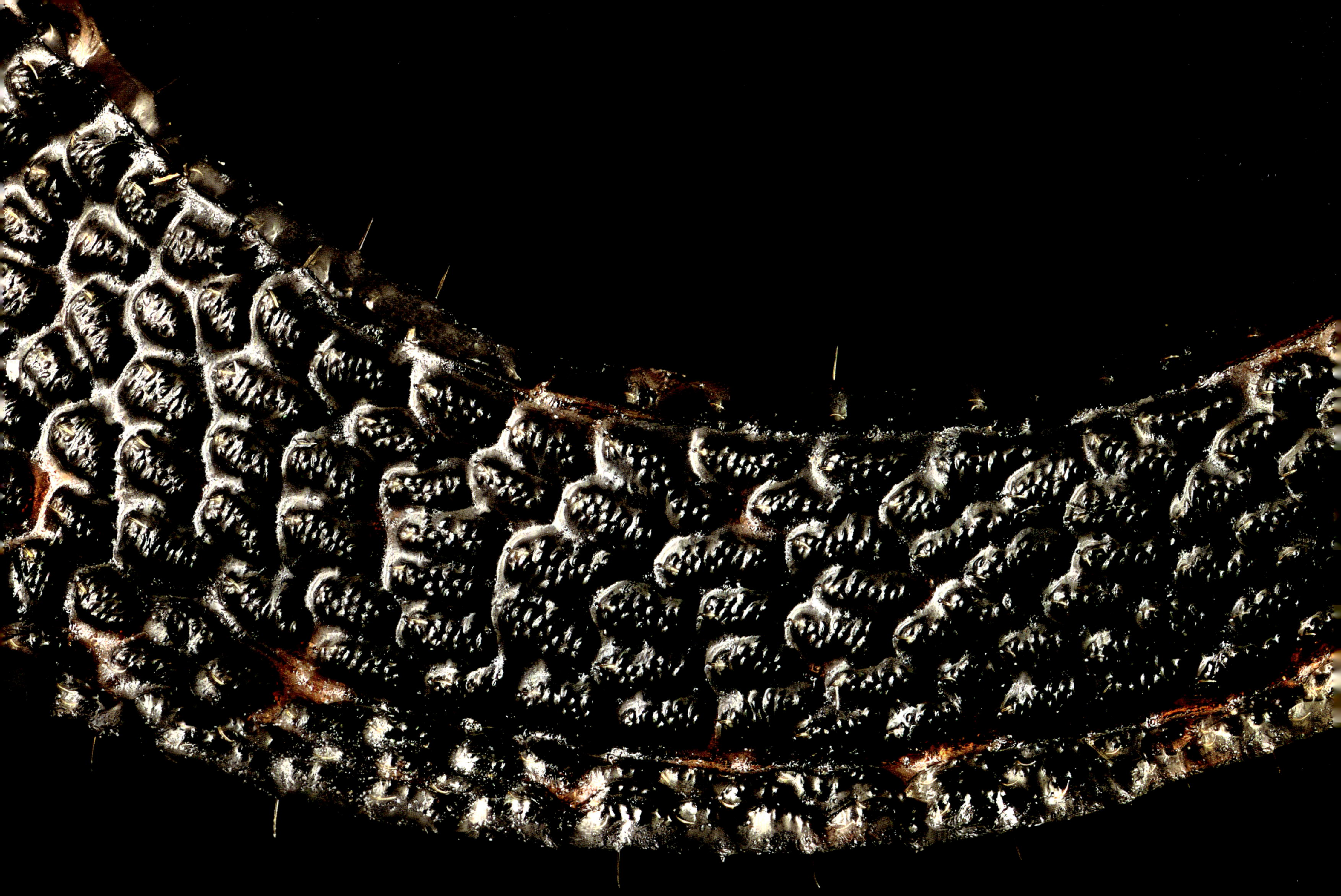

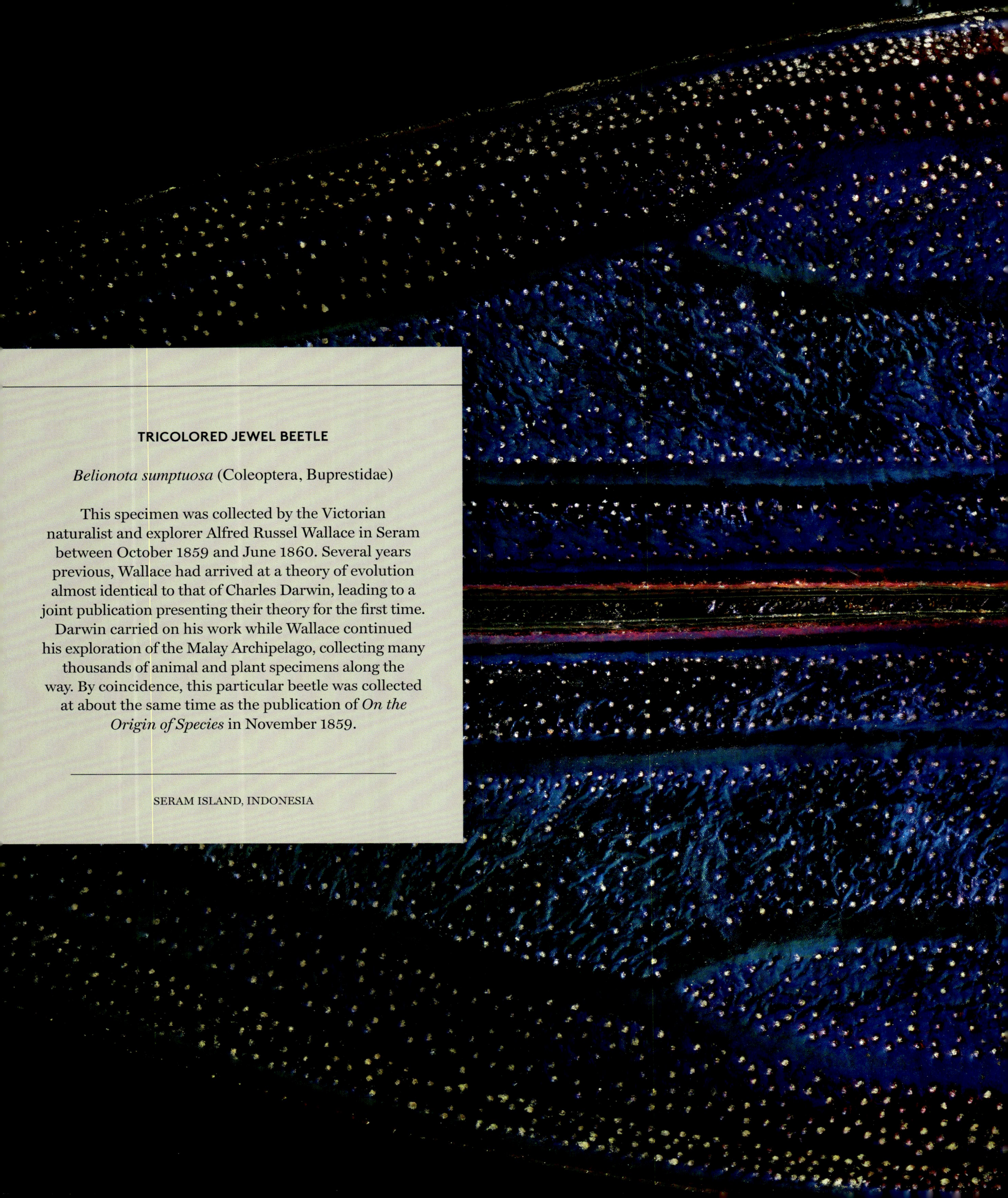

TRICOLORED JEWEL BEETLE

Belionota sumptuosa (Coleoptera, Buprestidae)

This specimen was collected by the Victorian naturalist and explorer Alfred Russel Wallace in Seram between October 1859 and June 1860. Several years previous, Wallace had arrived at a theory of evolution almost identical to that of Charles Darwin, leading to a joint publication presenting their theory for the first time. Darwin carried on his work while Wallace continued his exploration of the Malay Archipelago, collecting many thousands of animal and plant specimens along the way. By coincidence, this particular beetle was collected at about the same time as the publication of *On the Origin of Species* in November 1859.

SERAM ISLAND, INDONESIA

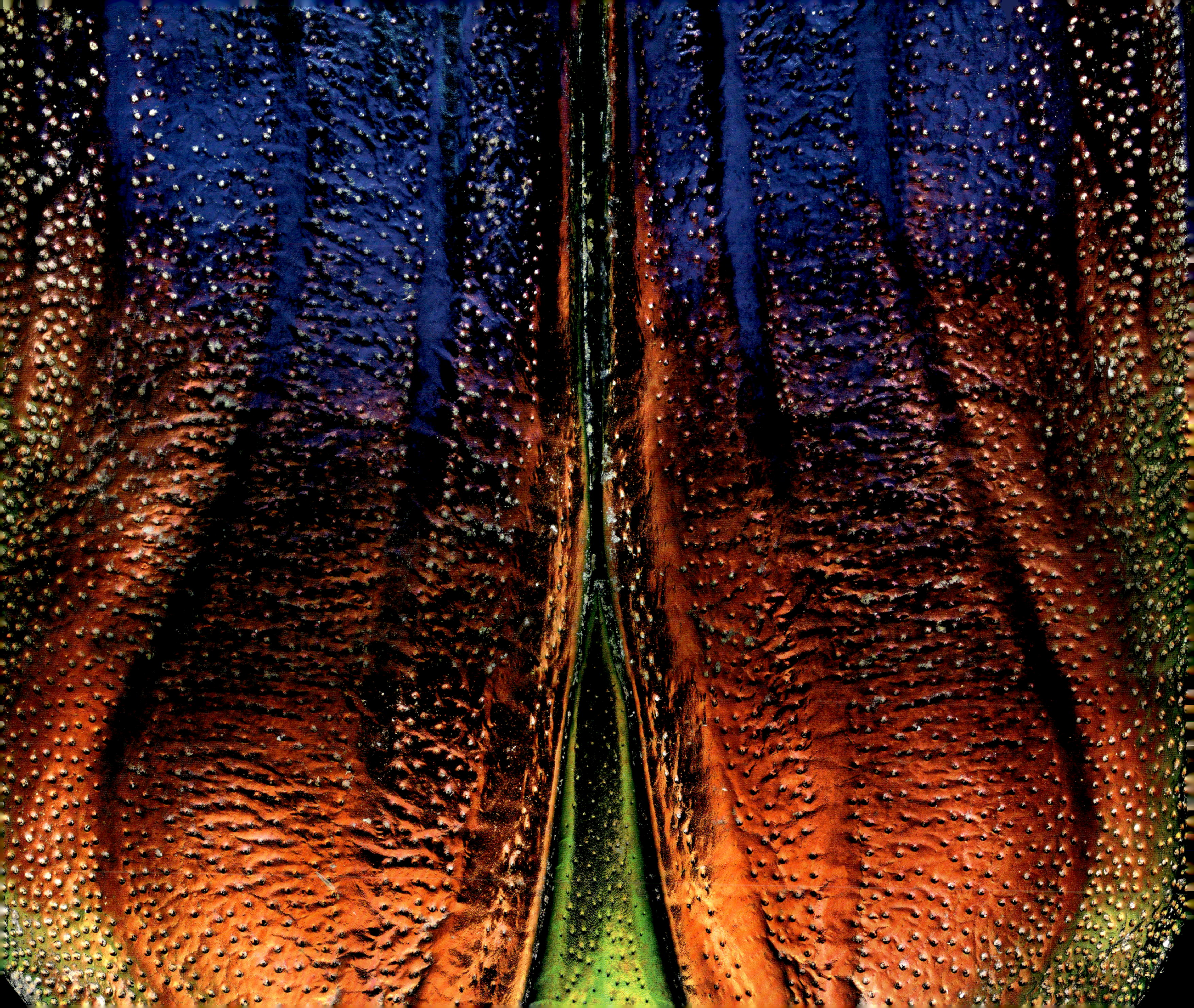

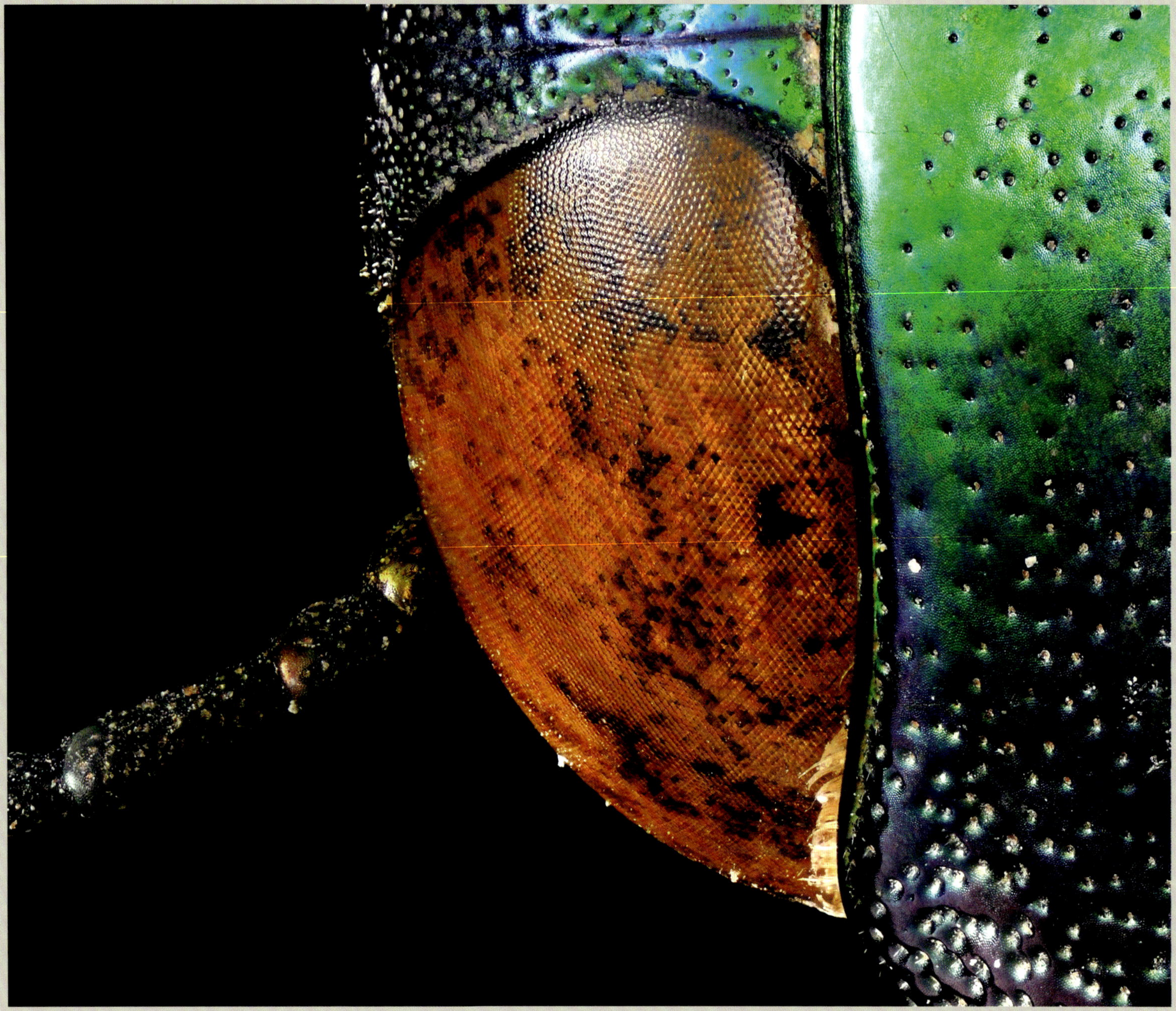

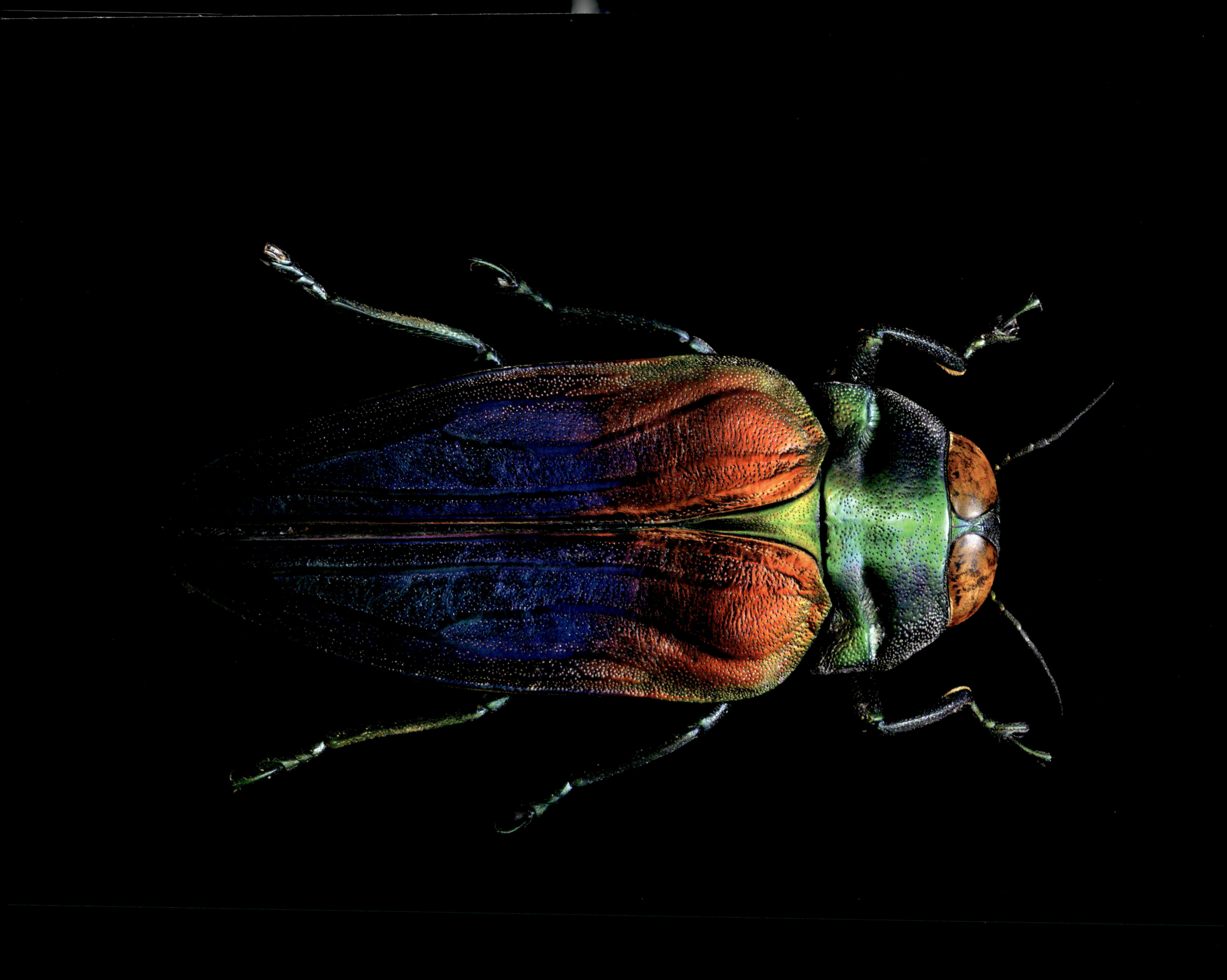

SPLENDID-NECKED DUNG BEETLE

Helictopleurus splendidicollis
(Coleoptera, Scarabaeidae)

The splendid-necked dung beetle is surely one of the most attractively marked dung beetles in the world. This species is just one of the many kinds unique to Madagascar. It is thought that after the extinction of all the large animals on Madagascar the splendid-necked dung beetle survived by switching from dung to carrion. Other species have survived by using alternative food sources such as cattle dung.

MADAGASCAR

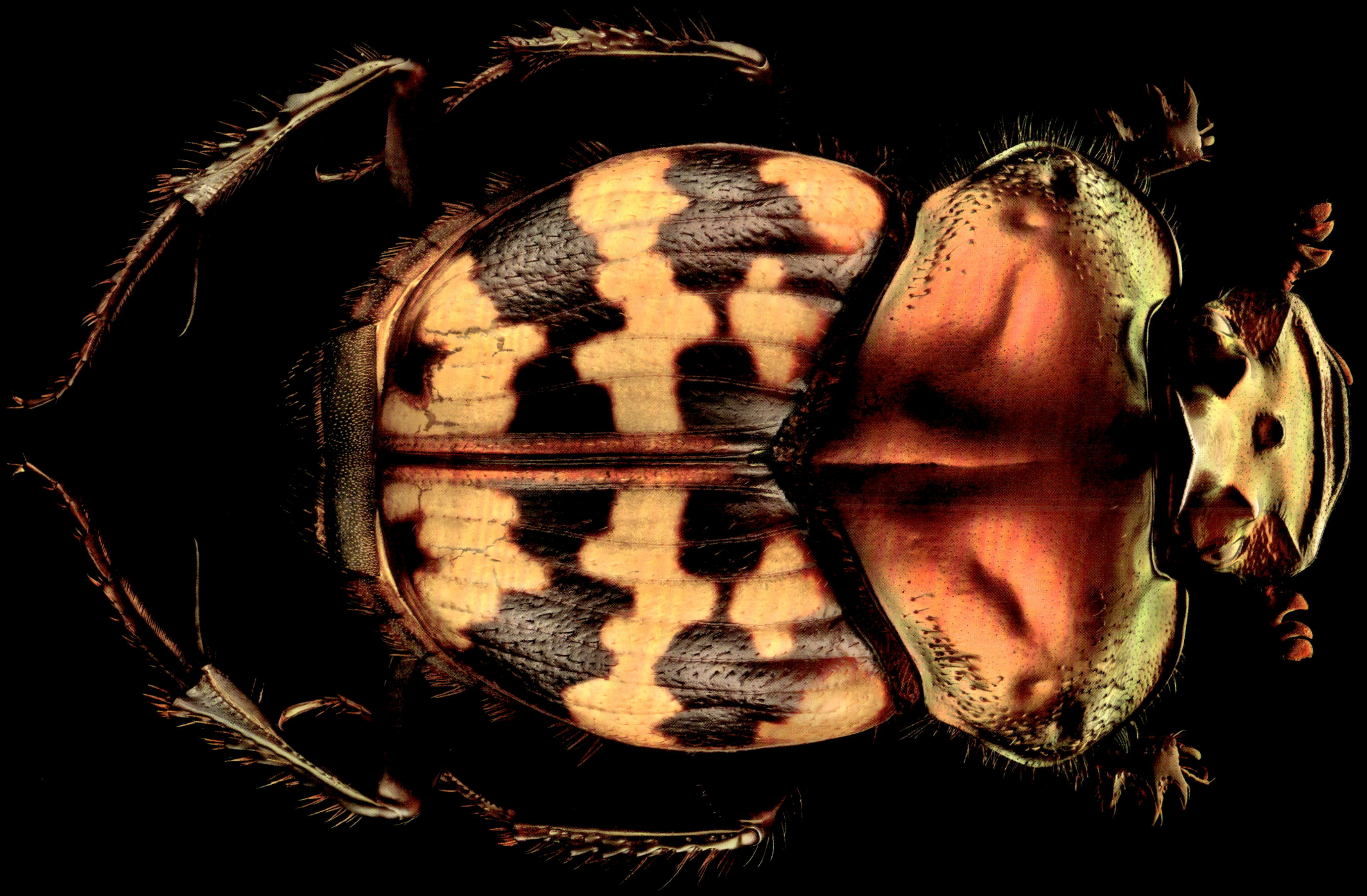

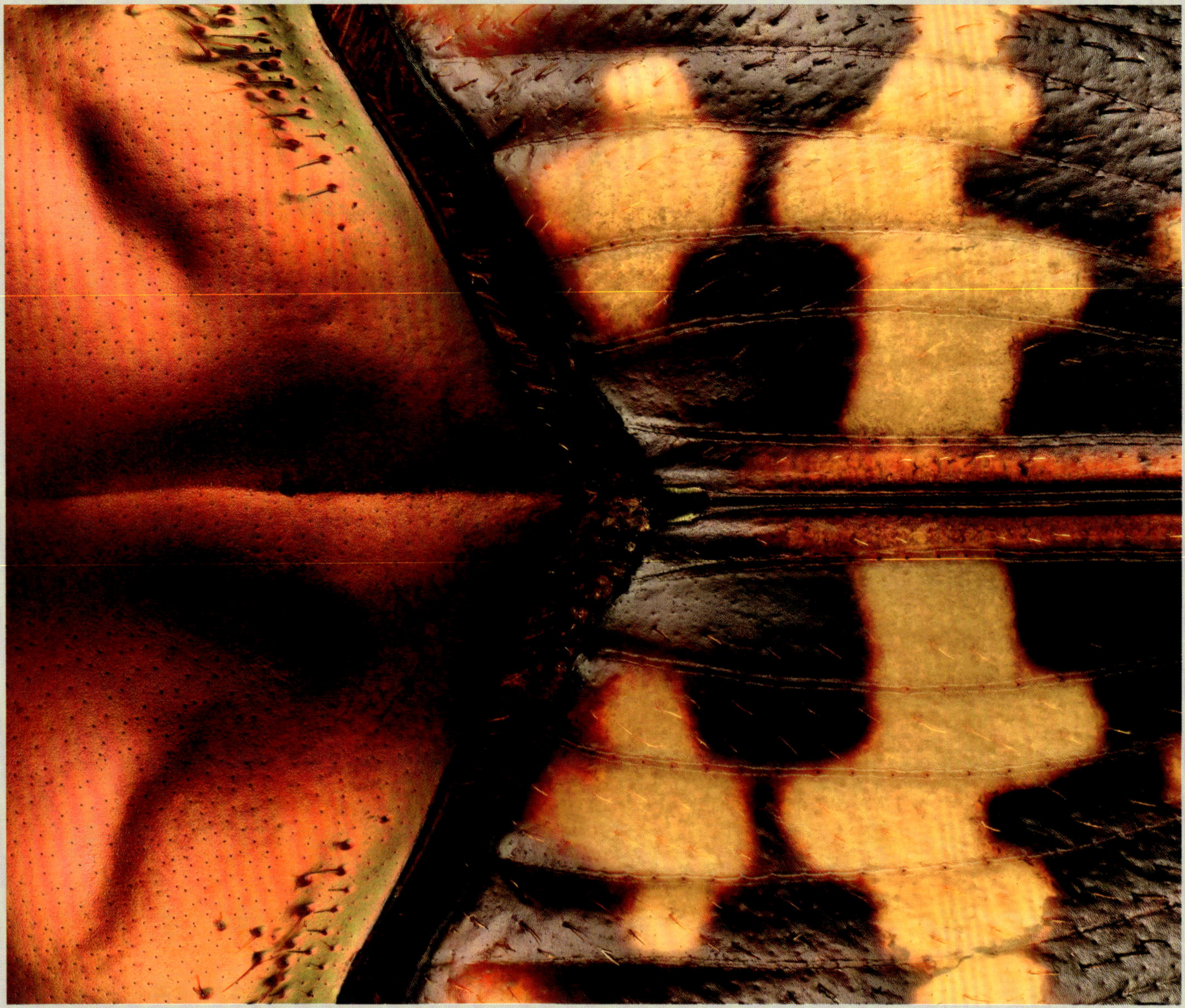

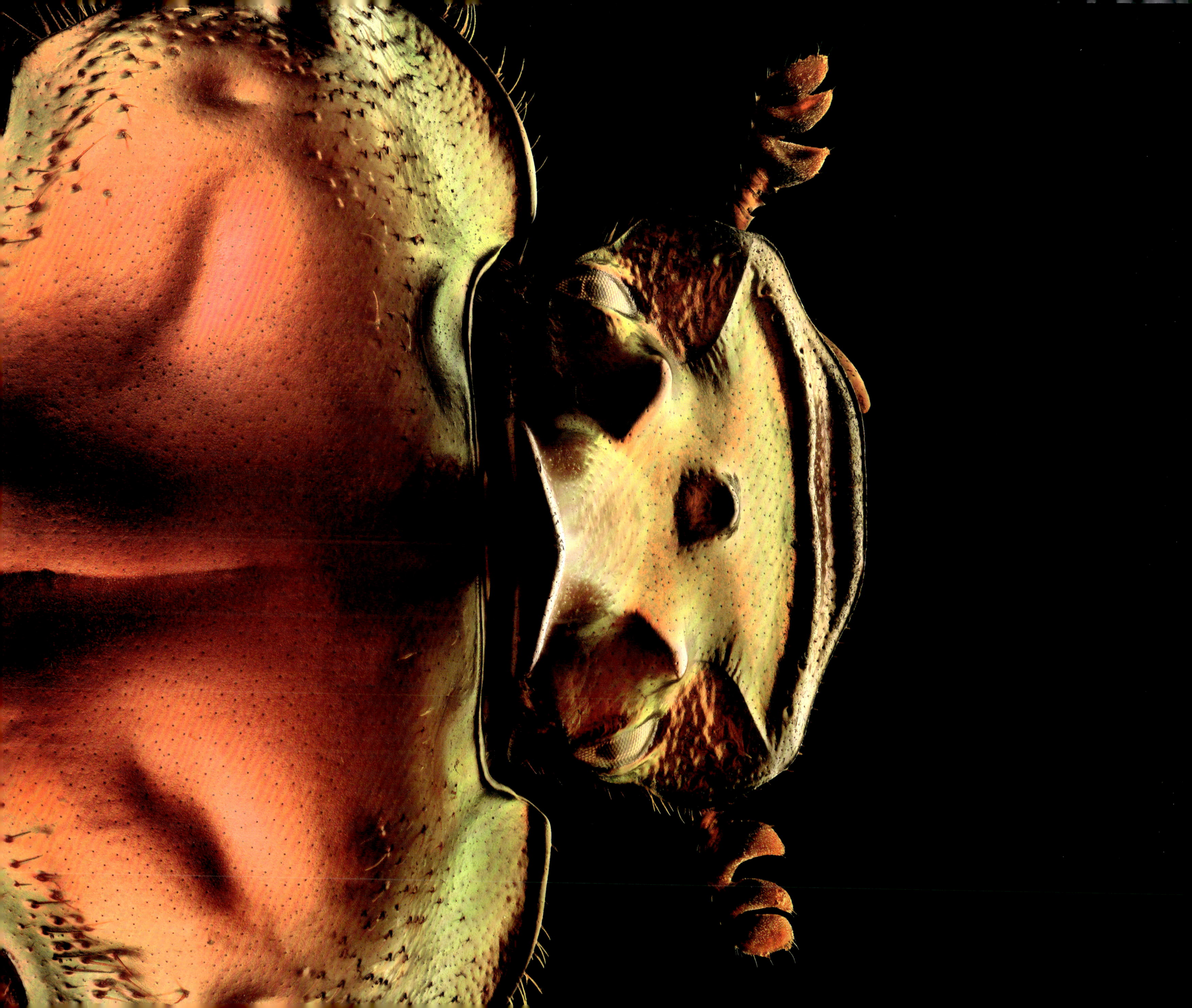

BLOWFLY

SPECIES OF THE FAMILY *Calliphoridae*
(Diptera, Calliphoridae)

Despite being unwelcome household visitors, these flies nevertheless perform essential roles in nature. Many blowflies help to recycle organic matter such as carrion and dung. Because of this, these flies are strongly attracted to the smell of decay, a fact which has been exploited by certain plants. By producing flowers with the scent of rotten meat, such plants recruit flies to serve as unwitting pollinators as they move from flower to flower.

ENGLAND

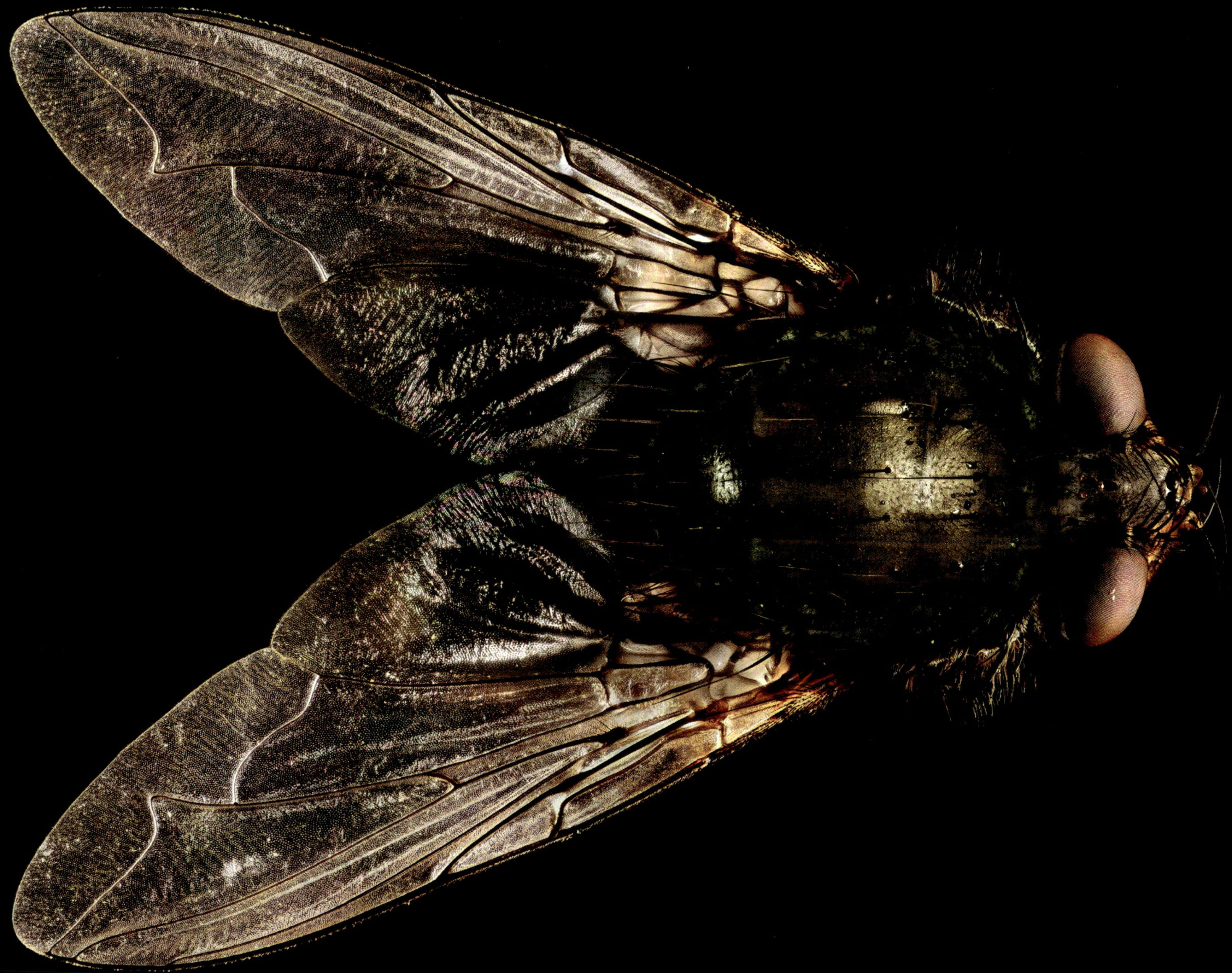

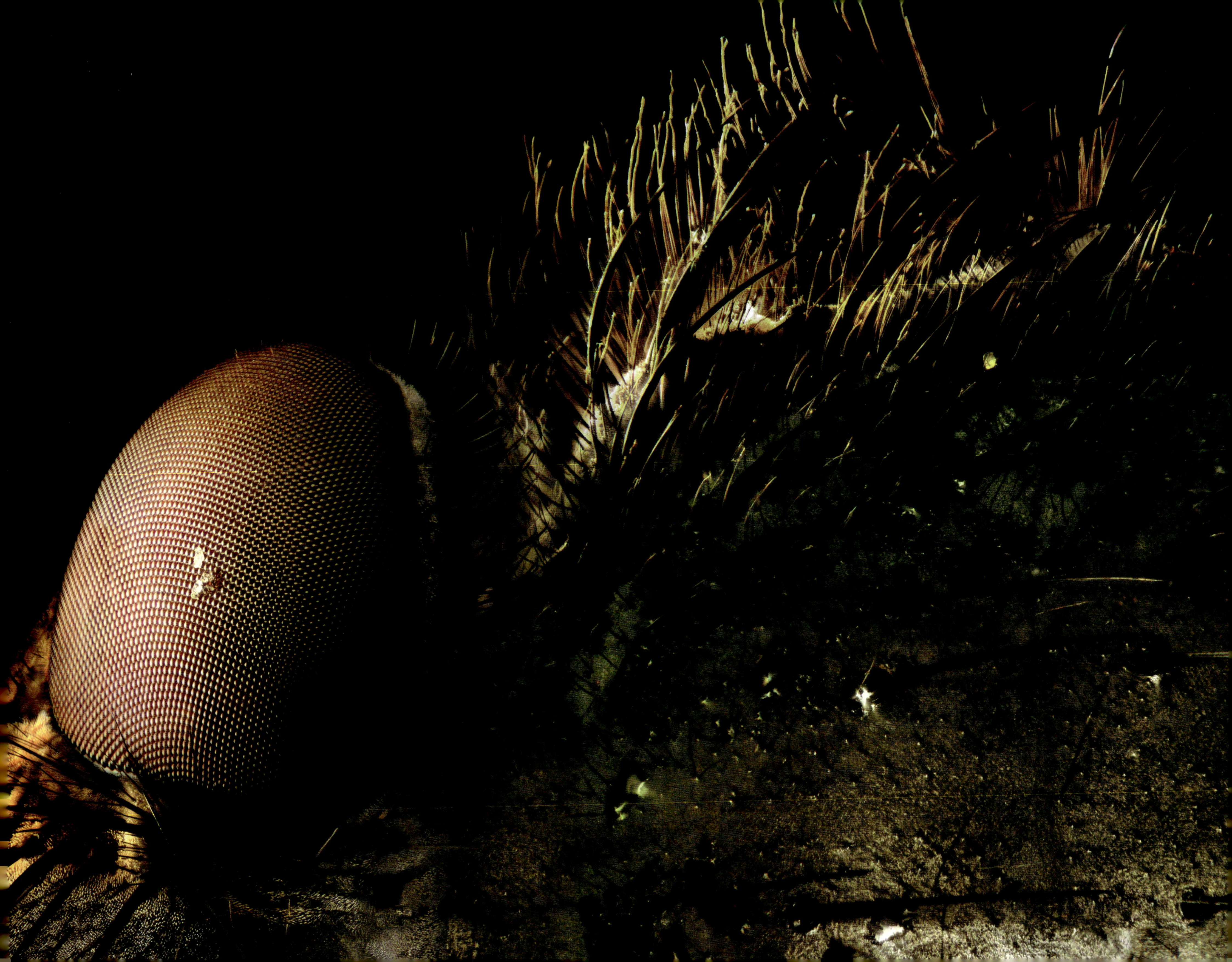

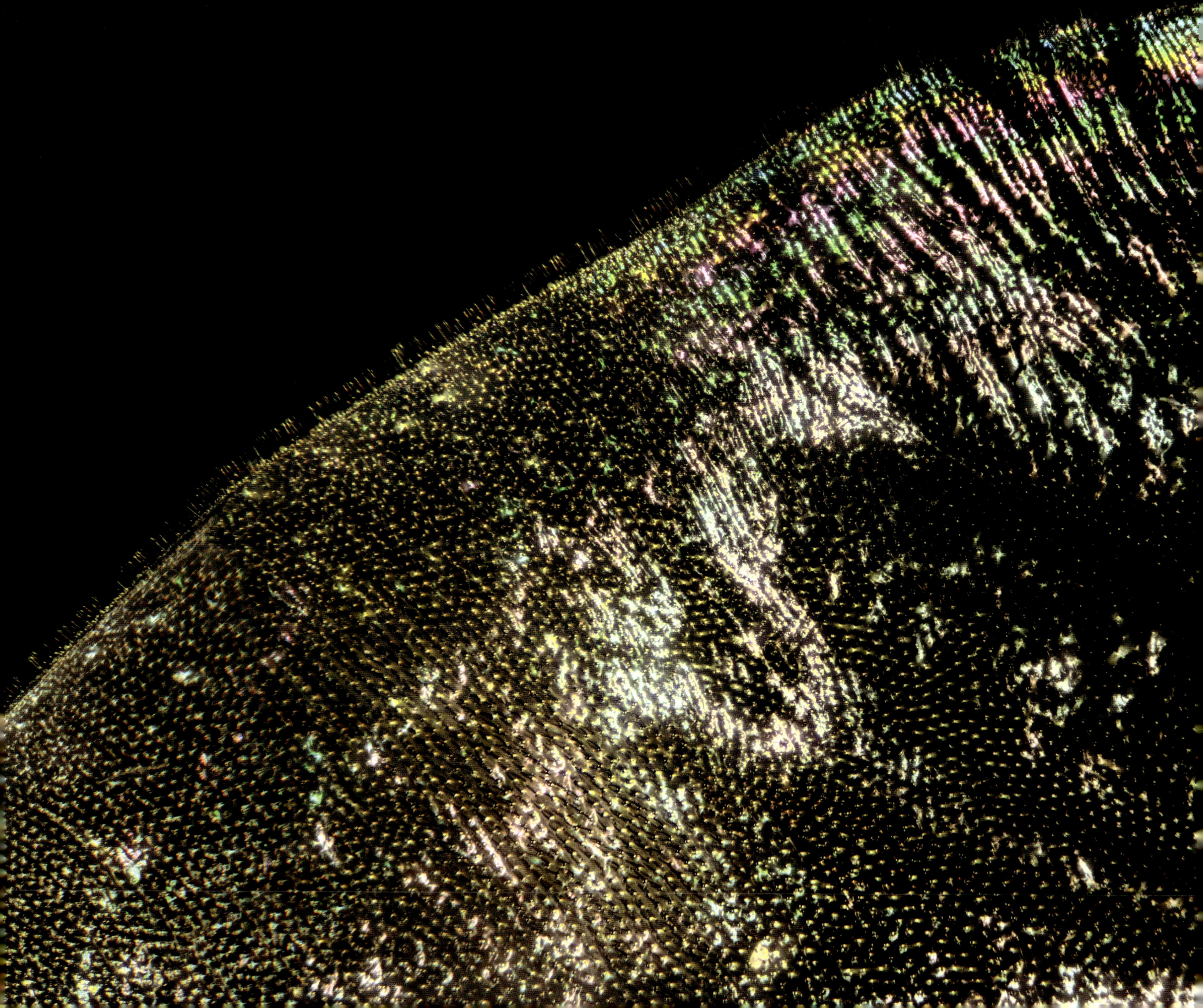

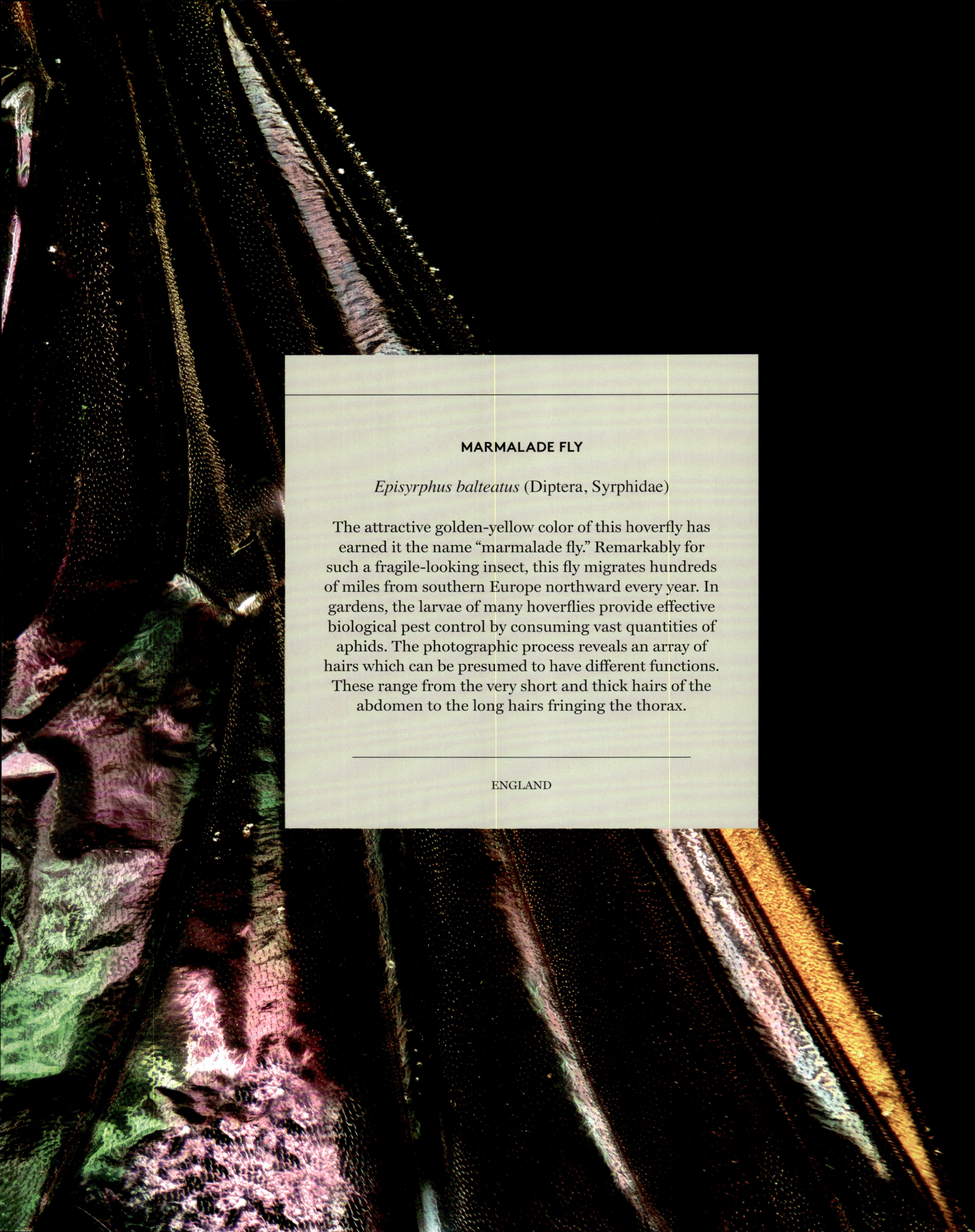

MARMALADE FLY

Episyrphus balteatus (Diptera, Syrphidae)

The attractive golden-yellow color of this hoverfly has earned it the name "marmalade fly." Remarkably for such a fragile-looking insect, this fly migrates hundreds of miles from southern Europe northward every year. In gardens, the larvae of many hoverflies provide effective biological pest control by consuming vast quantities of aphids. The photographic process reveals an array of hairs which can be presumed to have different functions. These range from the very short and thick hairs of the abdomen to the long hairs fringing the thorax.

ENGLAND

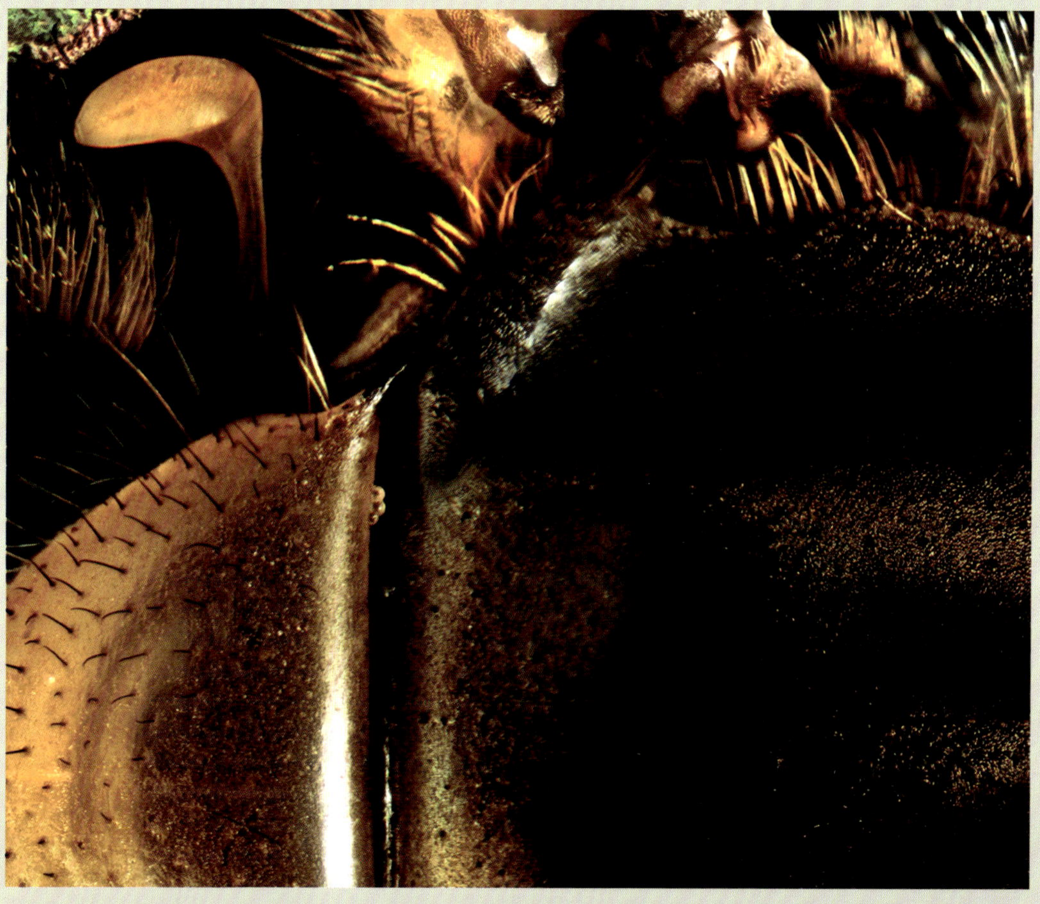
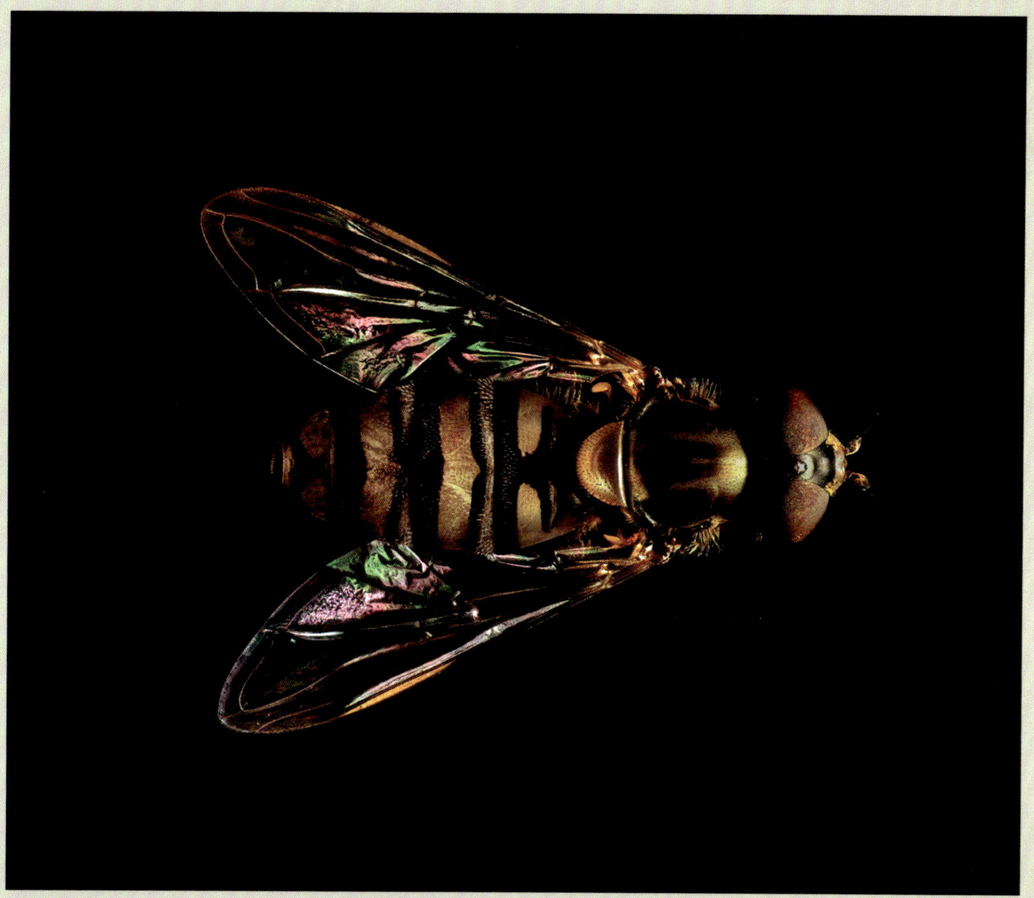
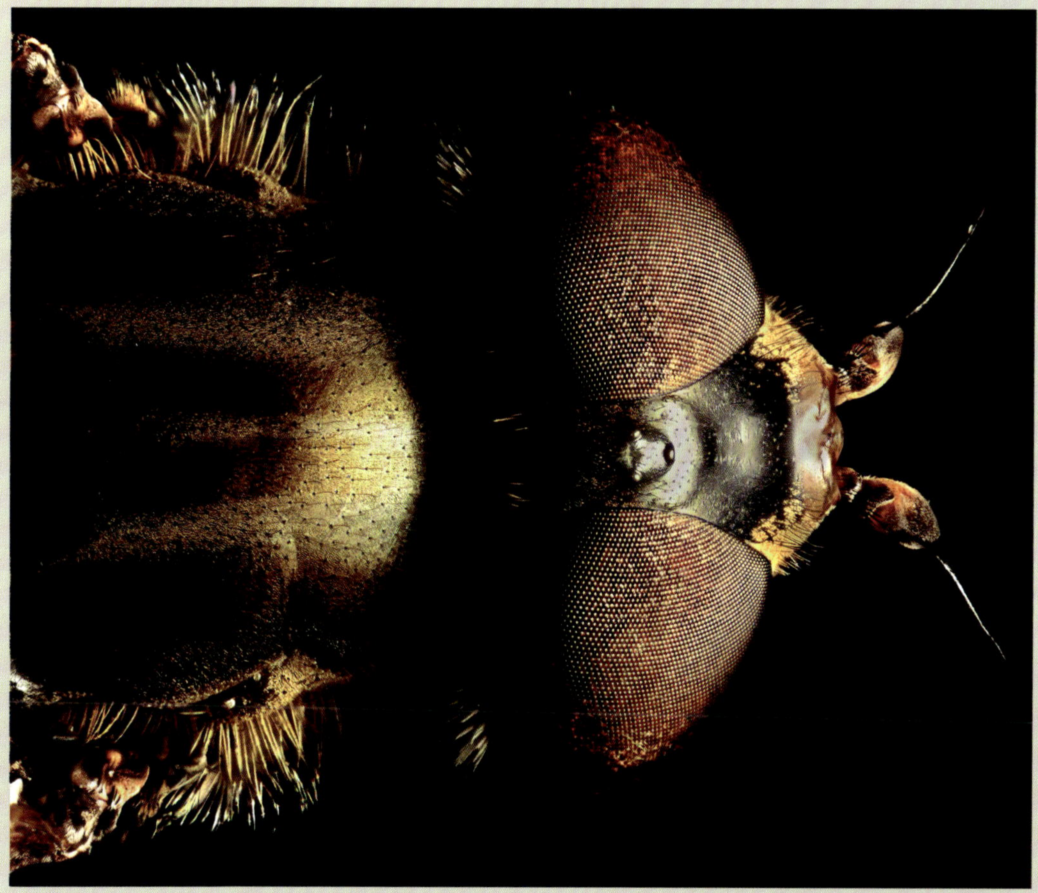
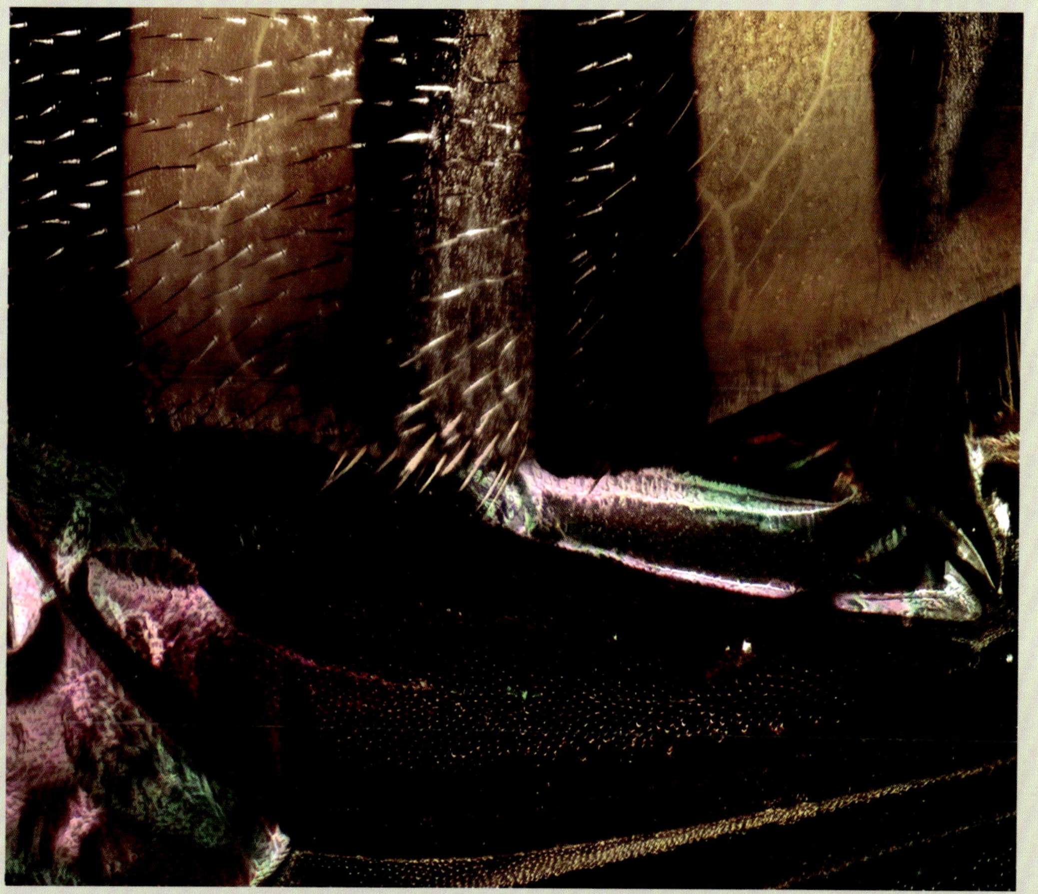

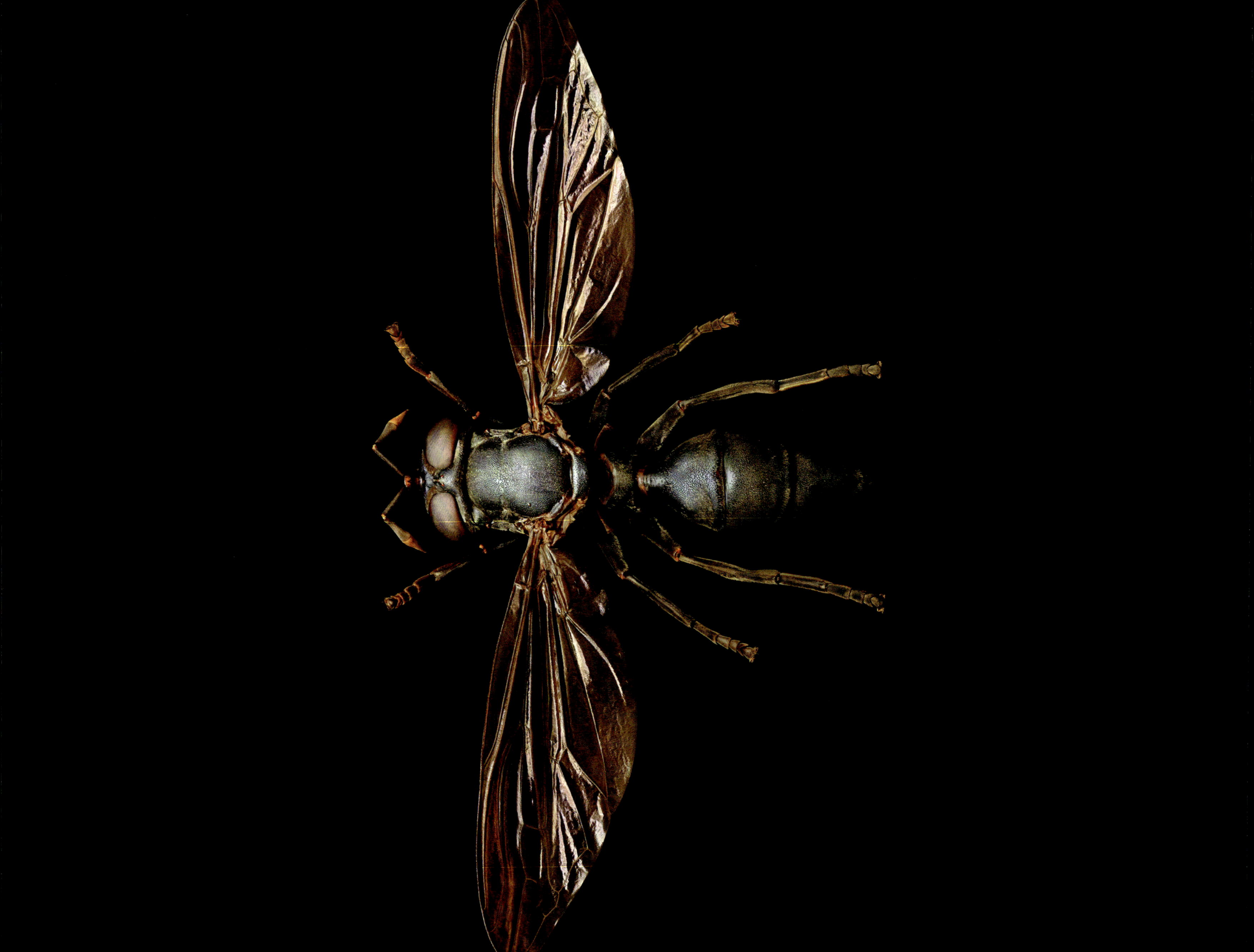

WASP MIMIC HOVERFLY

SPECIES OF THE GENUS *Cerioides* (Diptera, Syrphidae)

In nature, mimicry occurs when one species evolves a resemblance to another species. It is a fascinating phenomenon very common in the insect world. Often a harmless species evolves a resemblance to a distasteful or harmful species to gain protection. In this particular case the harmless wasp mimic hoverfly gains protection by mimicking aggressive and painfully stinging warrior wasps (species of *Synoeca*). The process of natural selection has produced a very accurate copy of the wasp, so that the fly has acquired dark wings, a narrow waist, longer antennae, and a reduction in body hairs.

BRAZIL

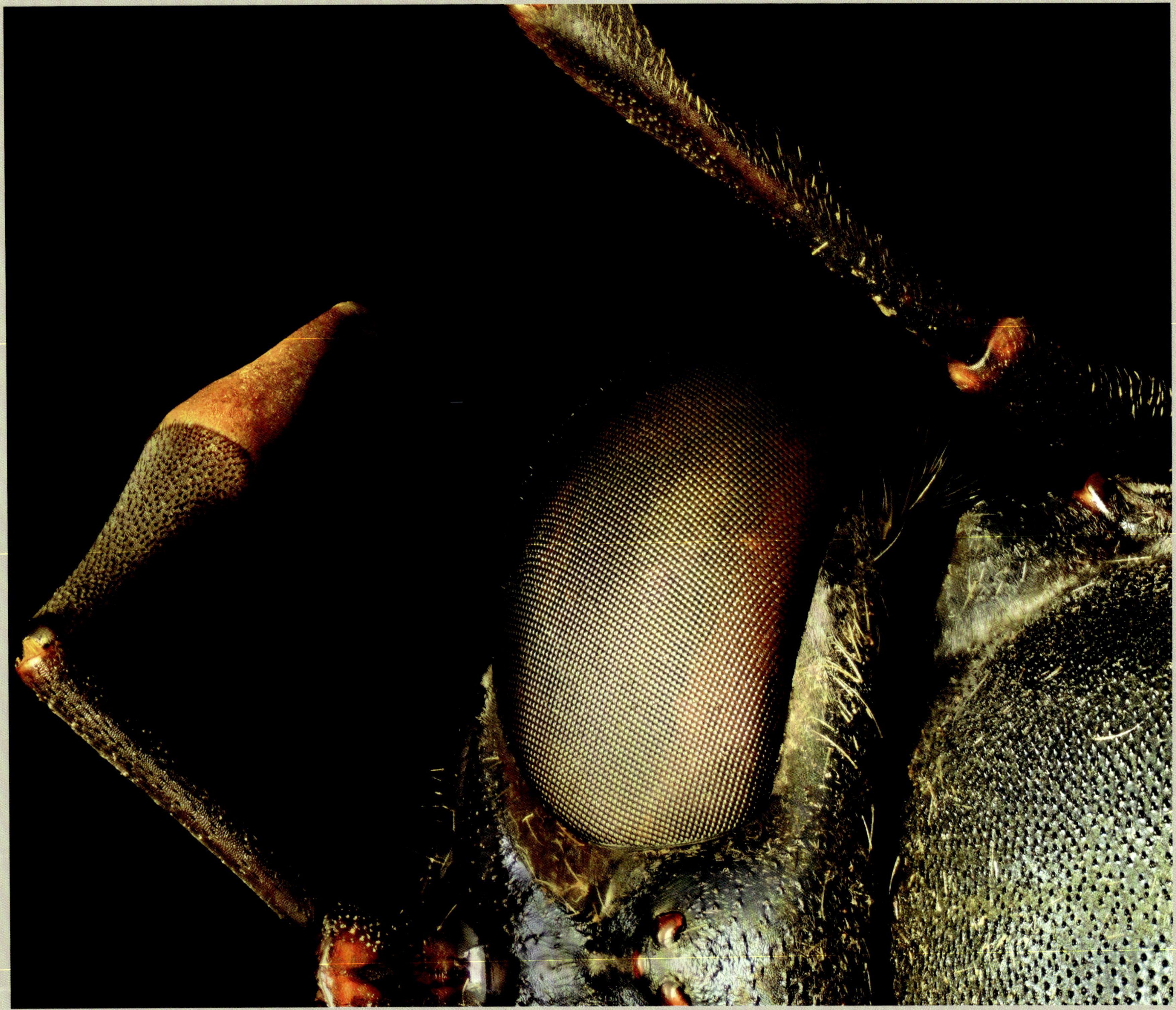

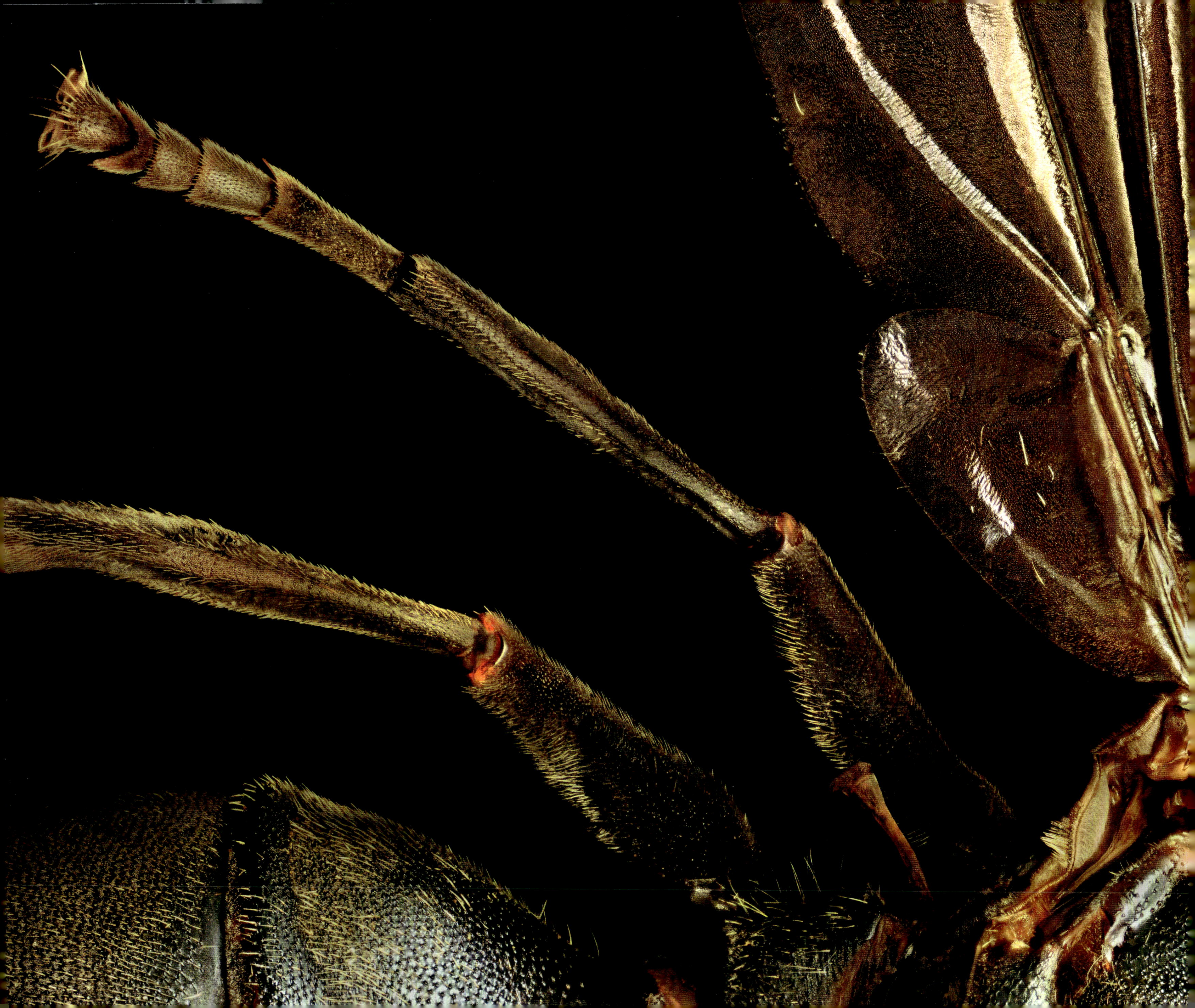

MANTIS-FLY

SPECIES OF THE GENUS *Mantispa*
(Neuroptera, Mantispidae)

Mantis-flies provide an excellent example of convergent evolution, where superficially similar features evolve in unrelated organisms. The heavily modified front legs used for seizing prey lend them a striking resemblance to praying mantises, but mantis-flies belong to a completely different group of insects: the lacewings (order Neuroptera). Similar grasping front legs have evolved independently several times in insects, in groups such as assassin bugs, flies, and thrips.

FRANCE

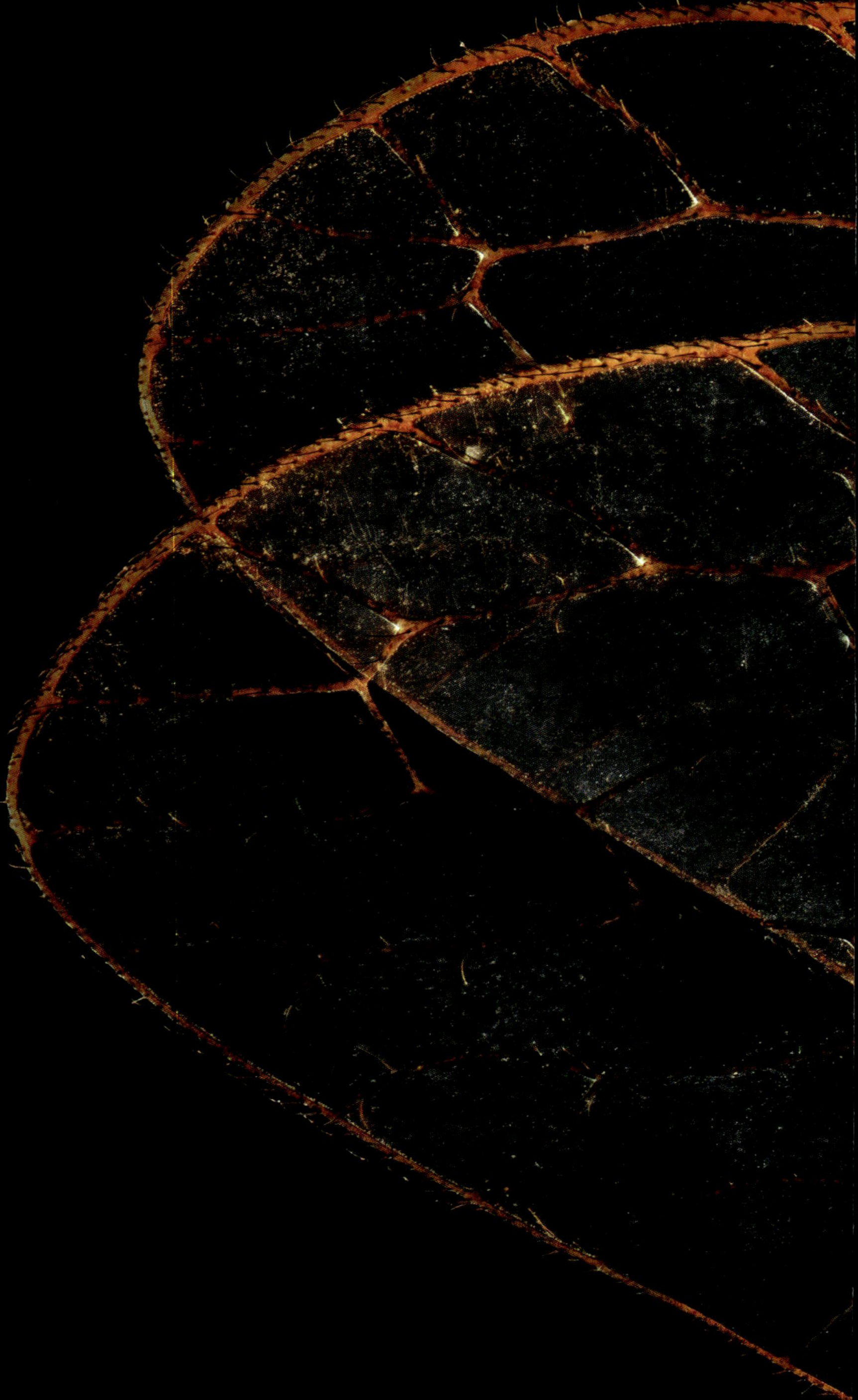

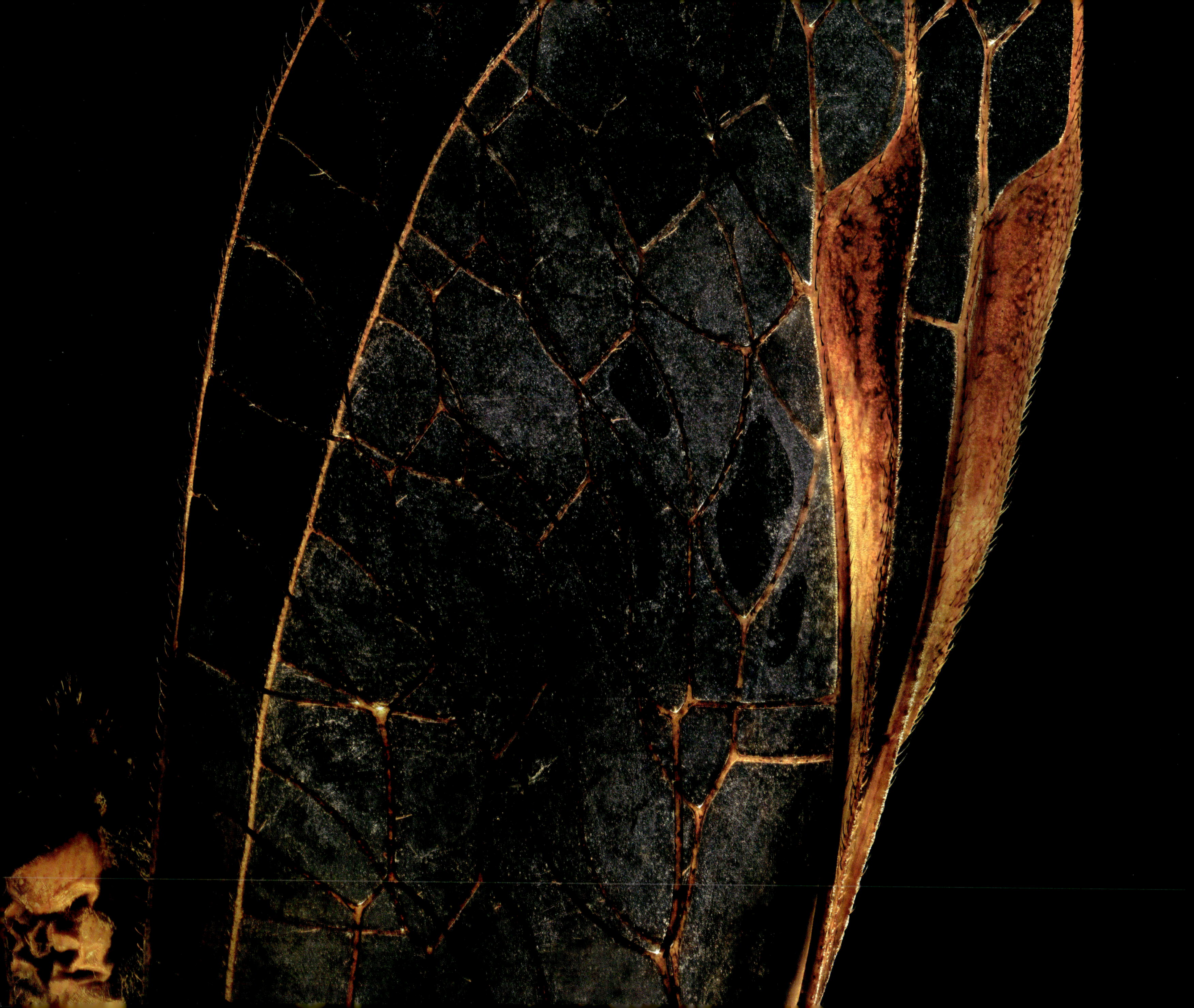

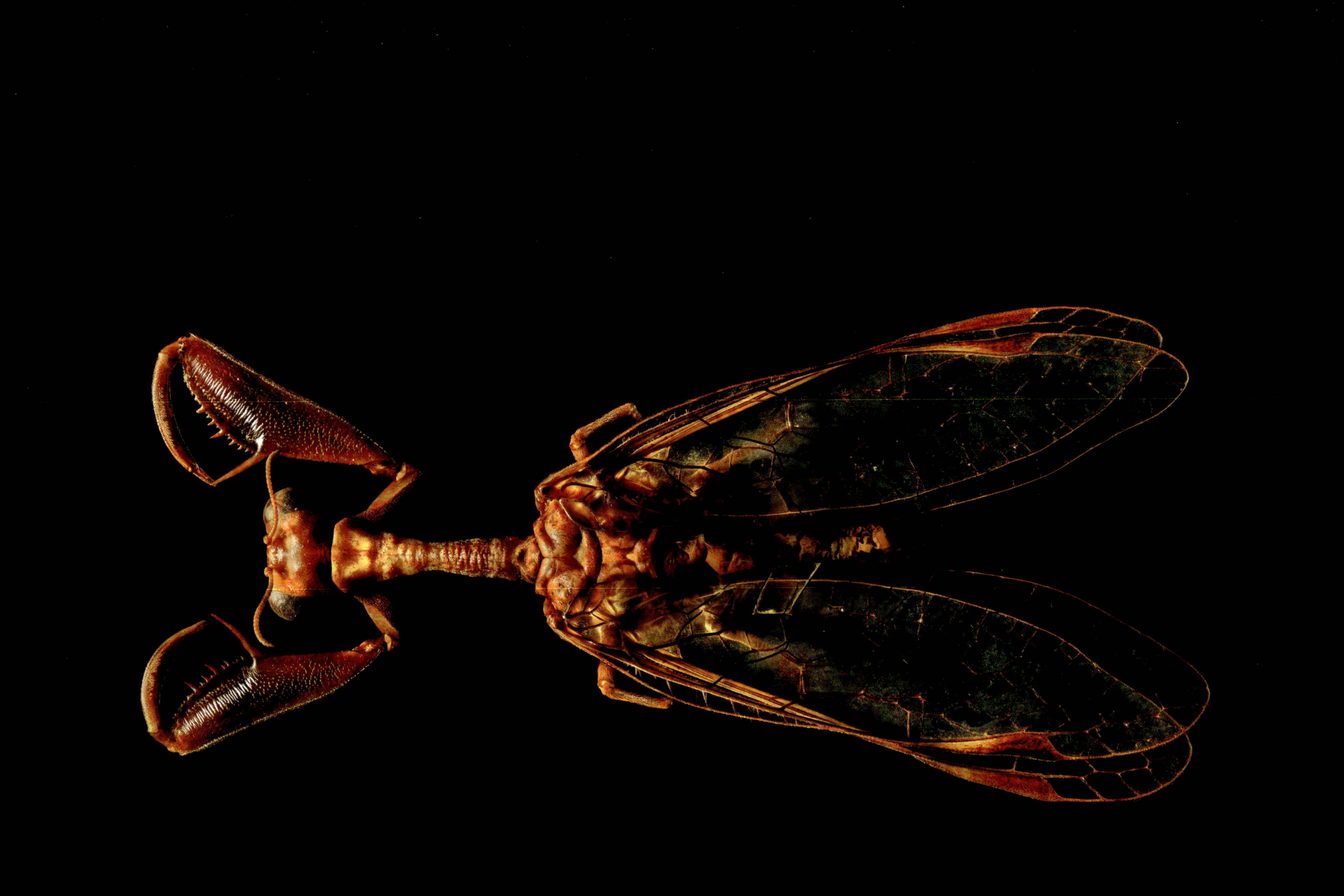

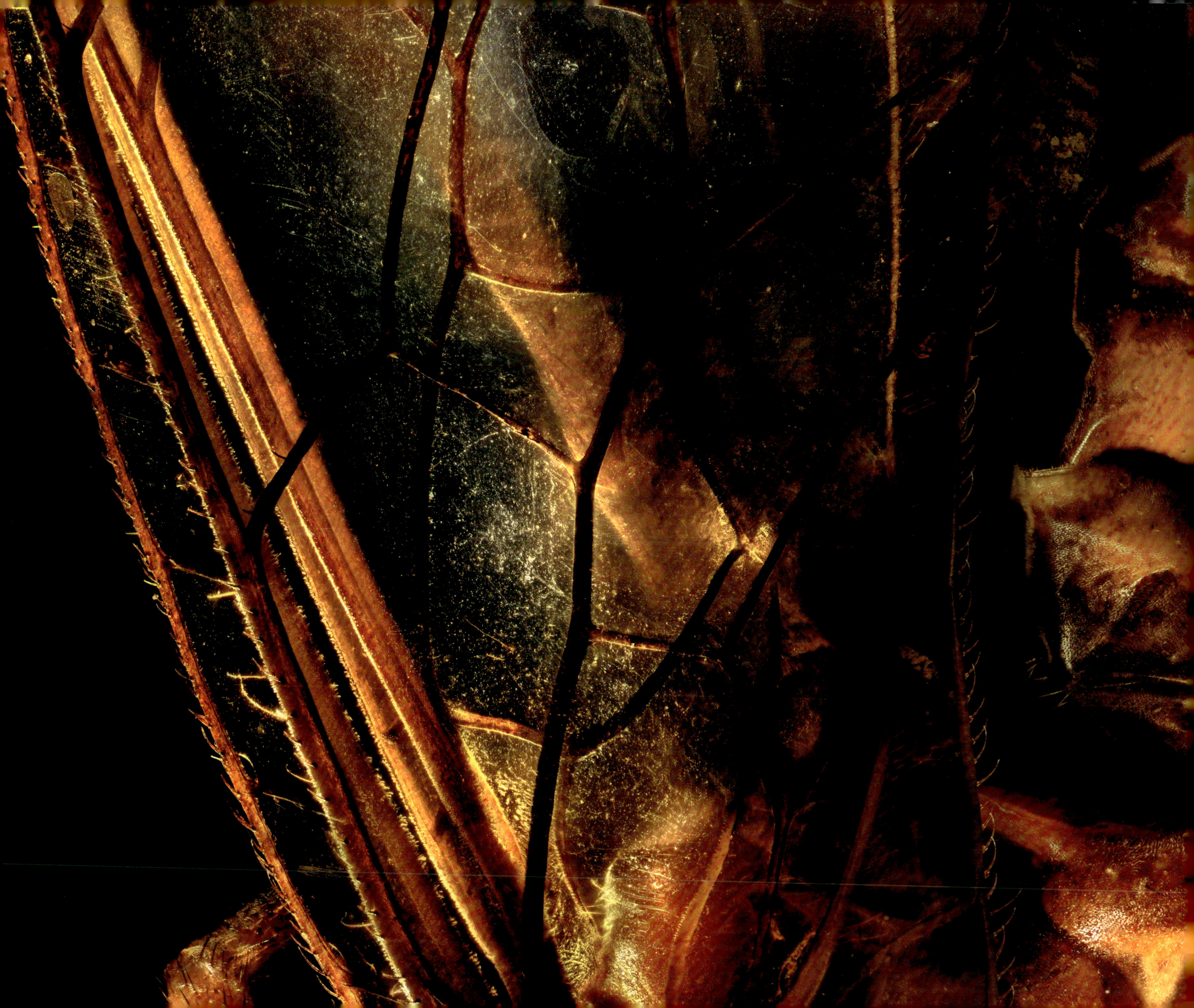

HAIRY-SIDED SEED EATER

Harpalus affinis (Coleoptera, Carabidae)

The majority of ground beetles are carnivorous, but this and related species are specialist seed eaters. By studying the structure of the mandibles it is possible to infer the diet of many ground beetles. In the case of the hairy-sided seed eater, the mandibles are short and blunt, adapted for breaking open seeds. The surface microsculpture of this beetle is particularly well developed and at high magnification the satin sheen of the wing cases is shown to be caused by a complex mesh-like covering.

ENGLAND

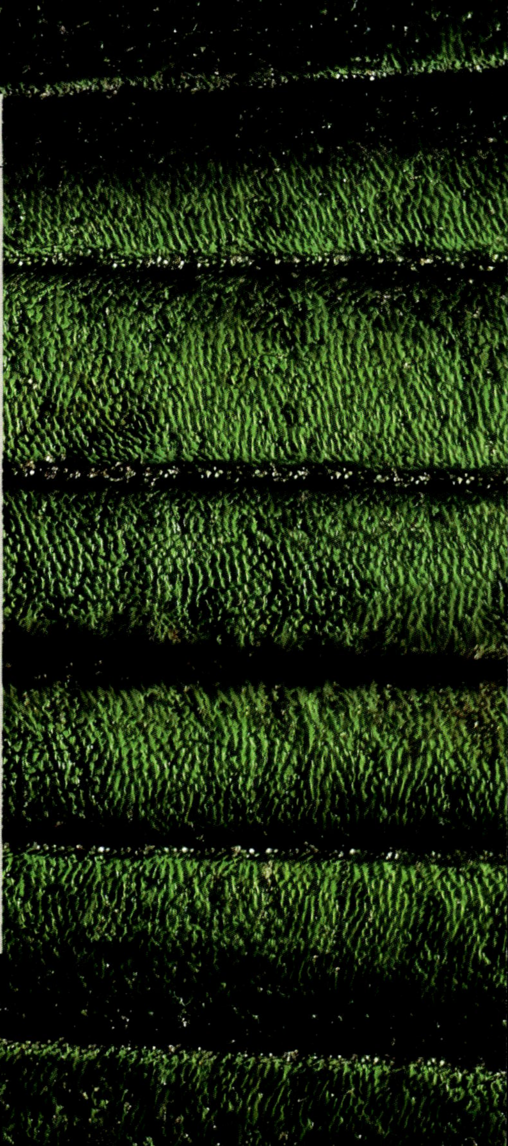

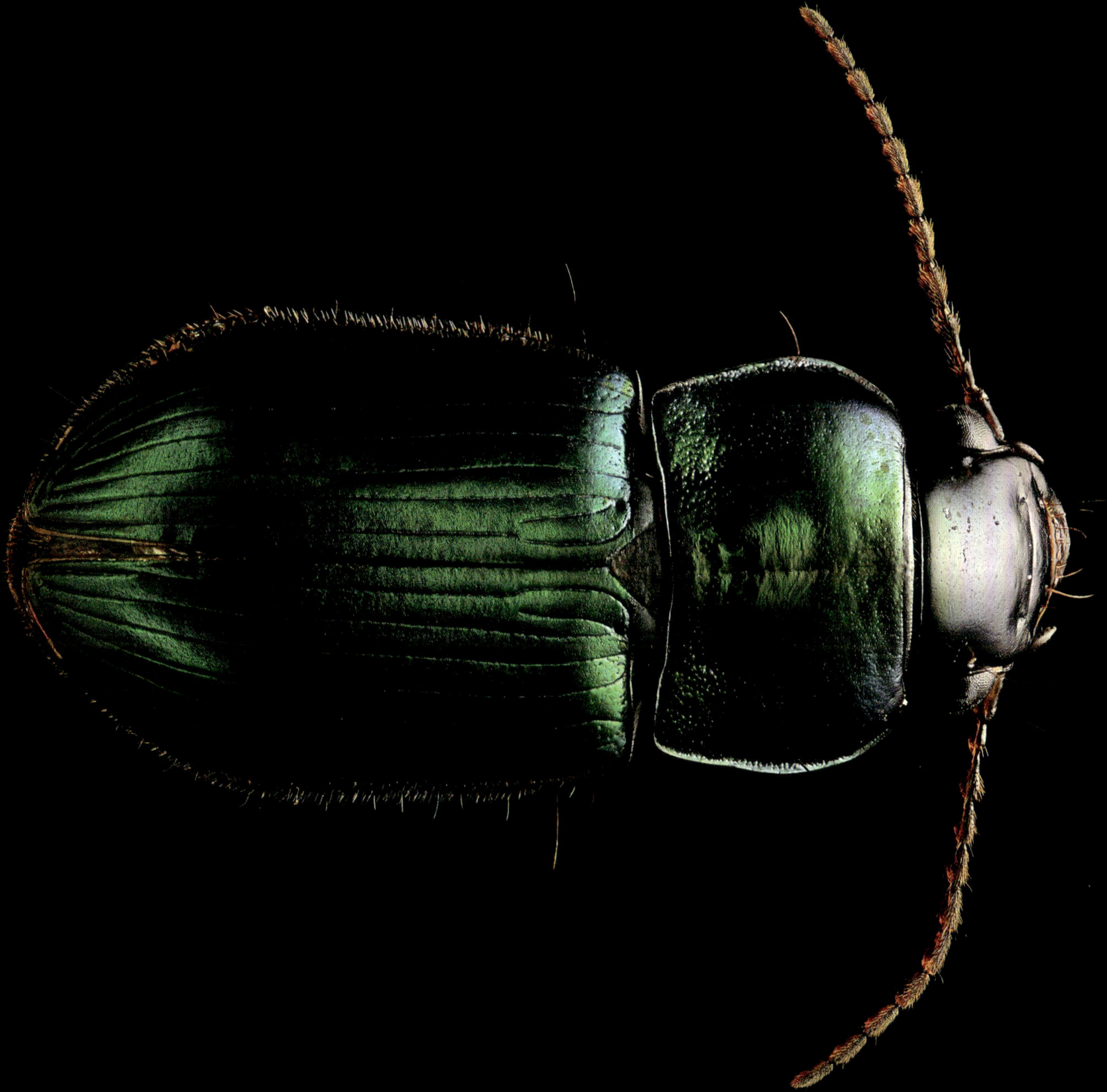

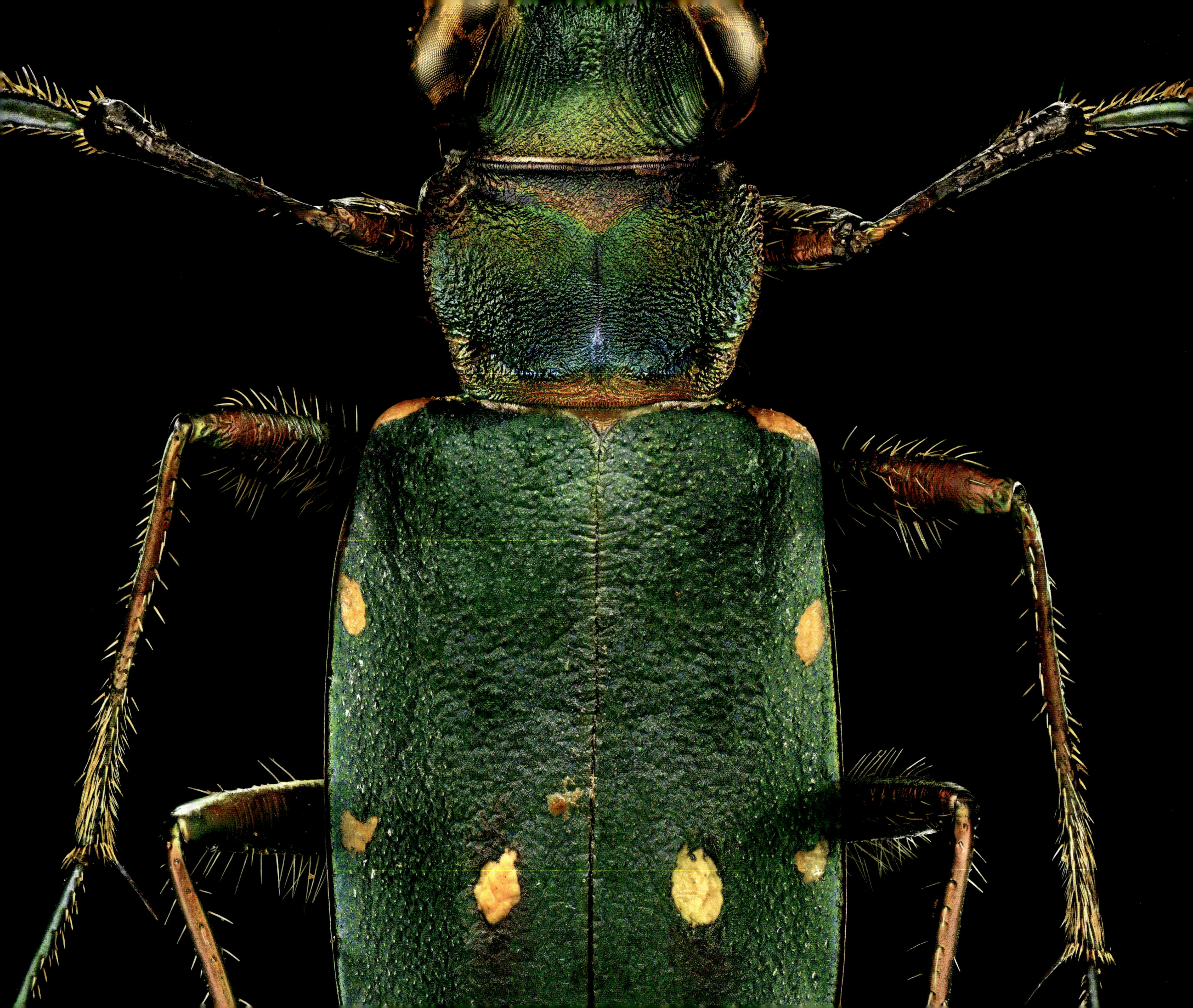

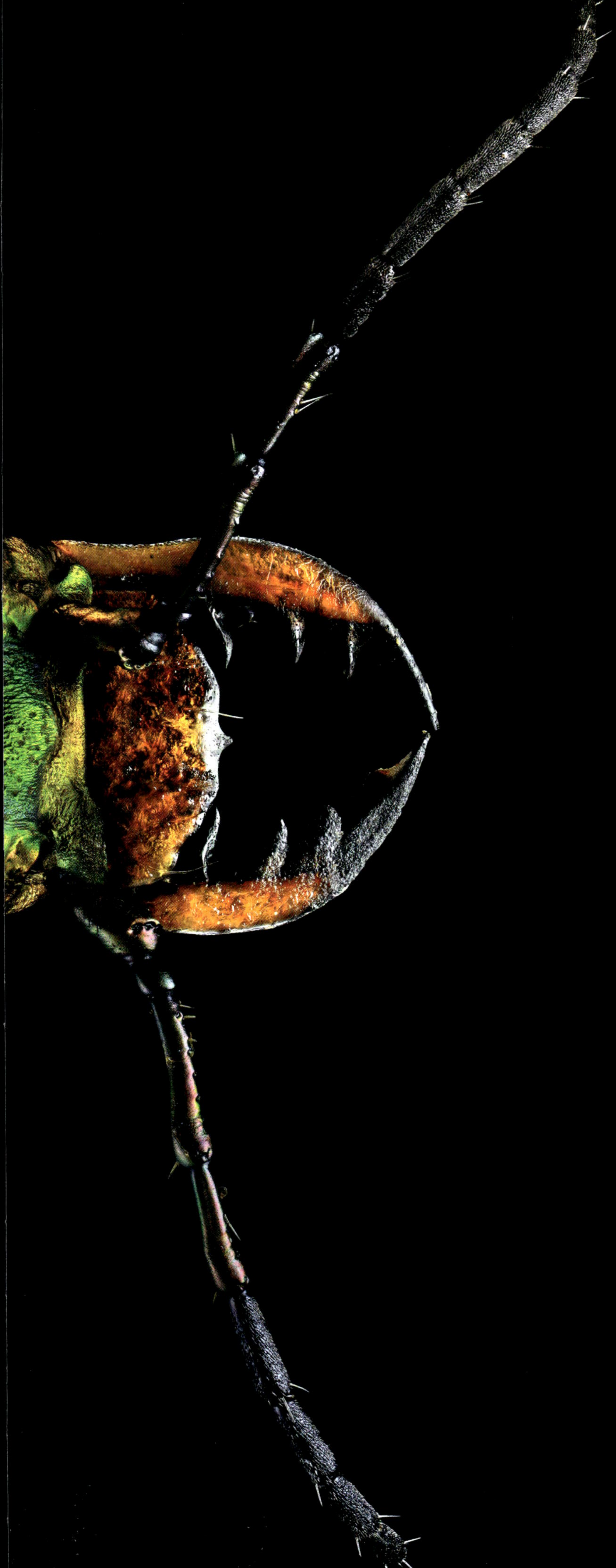

GREEN TIGER BEETLE

Cicindela campestris (Coleoptera, Carabidae)

In temperate regions tiger beetles are particularly common in sandy areas. They are fast-running, very alert insects, and, in warm weather, difficult to approach and observe. Often their presence is only revealed when the beetle takes to the air in a fleeting flash of brilliant color. While the adults are active hunters, the larvae are sedentary, lying in wait in burrows to ambush passing prey.

ENGLAND

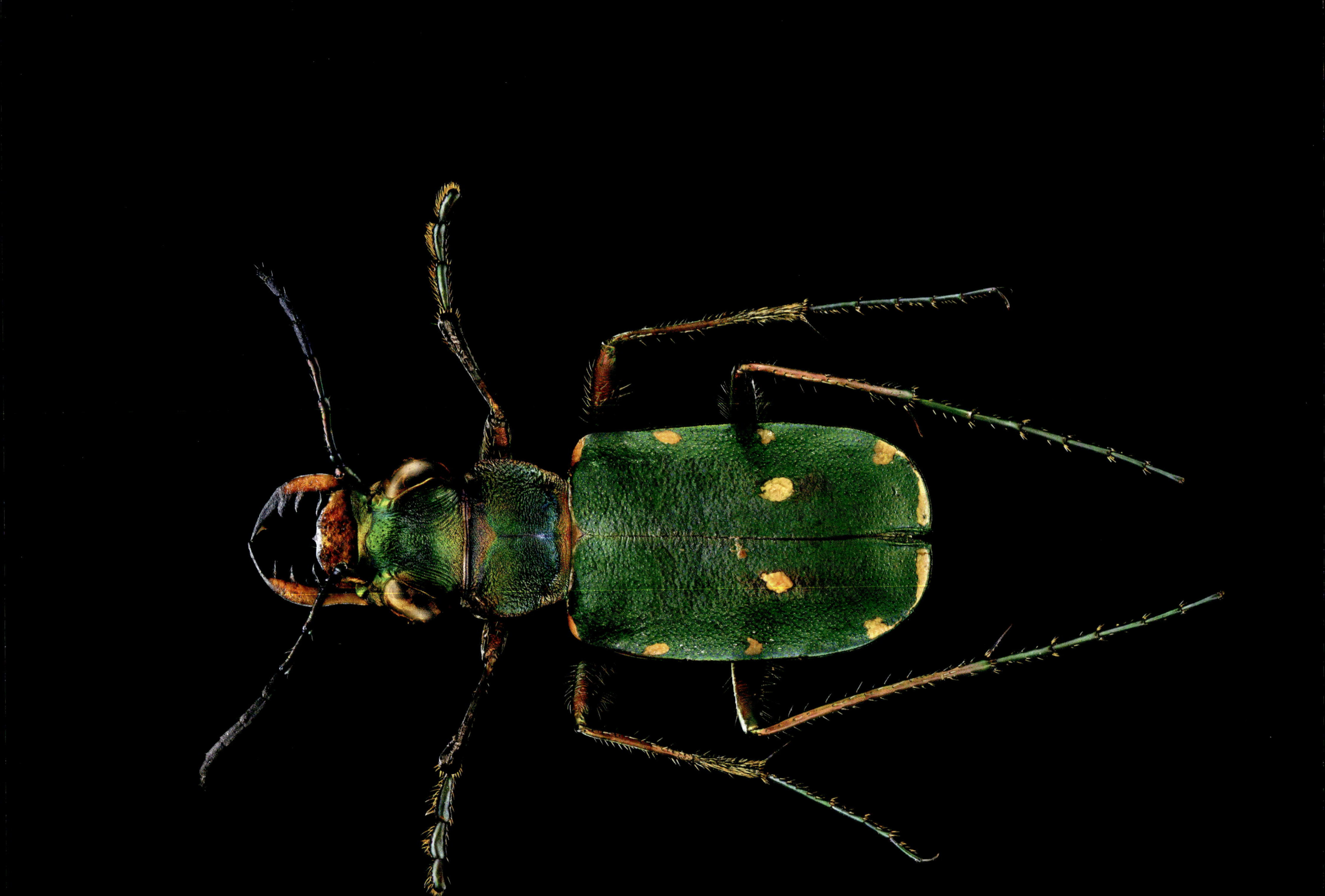

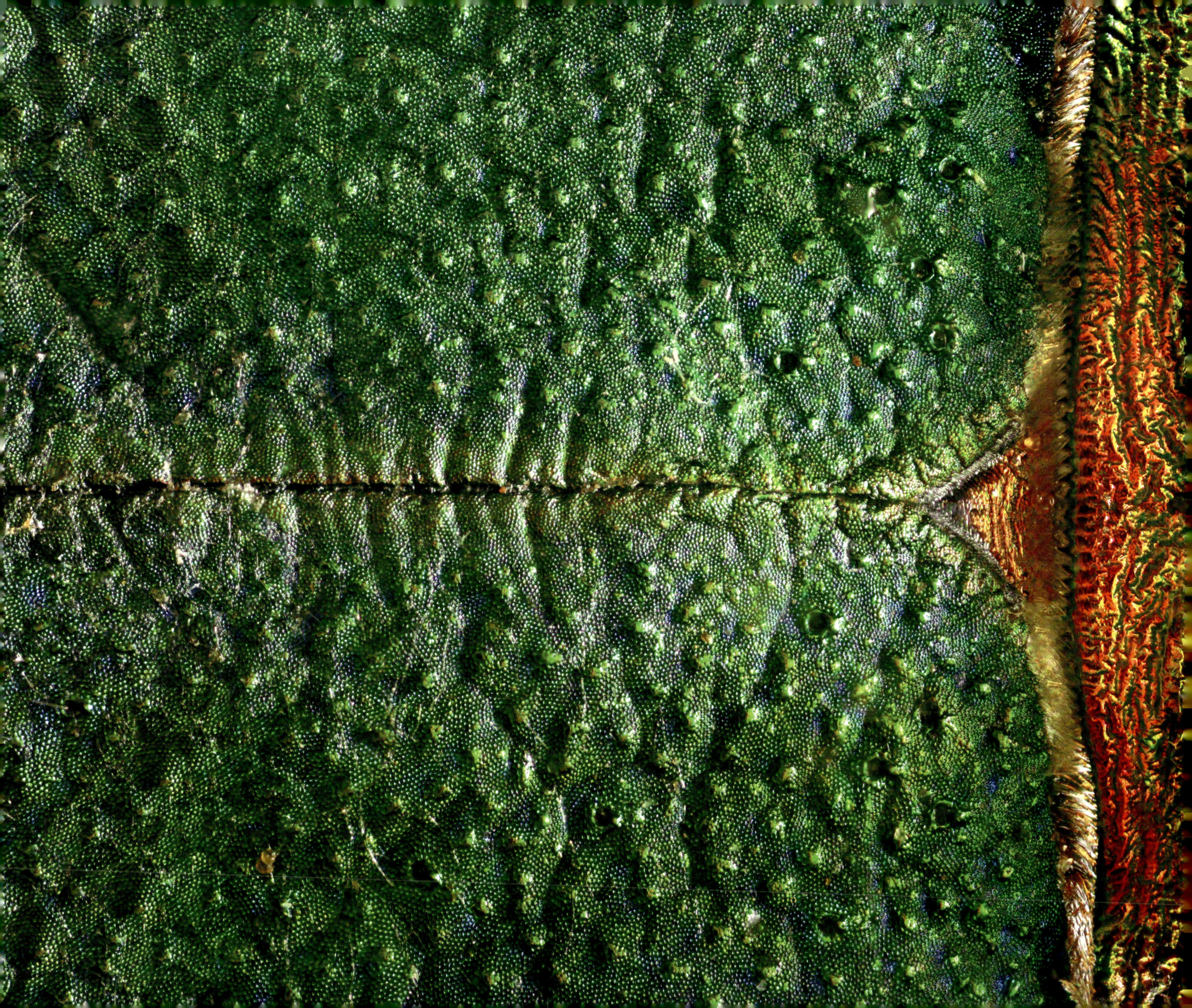

TORTOISE BEETLE

Platypria melli (Coleoptera, Chrysomelidae)

The domed body shape of tortoise beetles probably provides protection against ants and other small predators. Indeed, as they move over the surface of a leaf they are reminiscent of miniature tanks. The softer-bodied larvae employ a different method, constructing a defensive shield of their own feces and shed skin. The wonderfully complex shape of this particular tortoise beetle is a composite of bumps, pits, wrinkles, and rows of spines. Light passing through the specimen further reveals areas where the wing cases are paper-thin. The function of the spines is unclear, but can be assumed to be an anti-predator defense or help in camouflaging the insect on its host plant.

CHINA

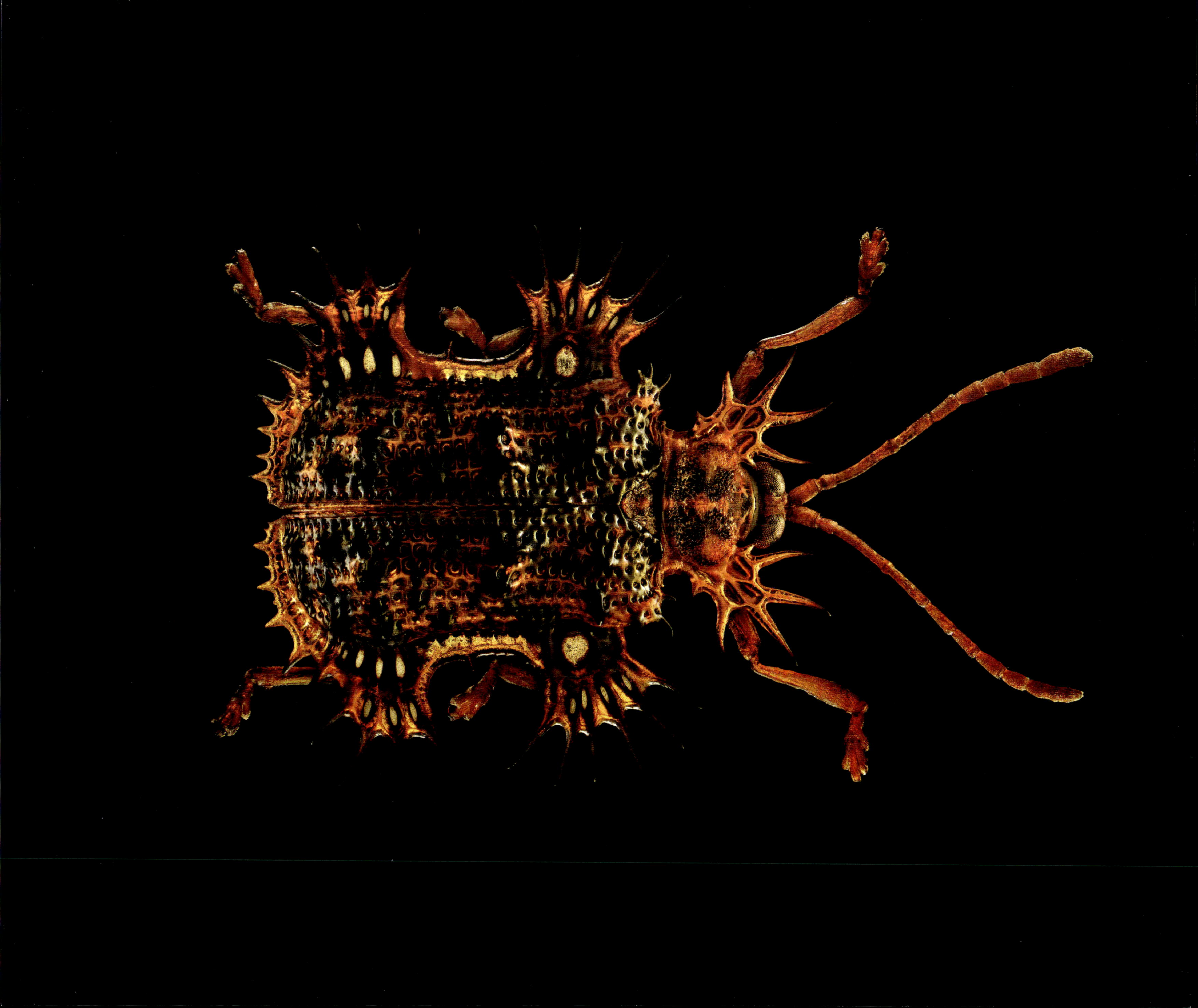

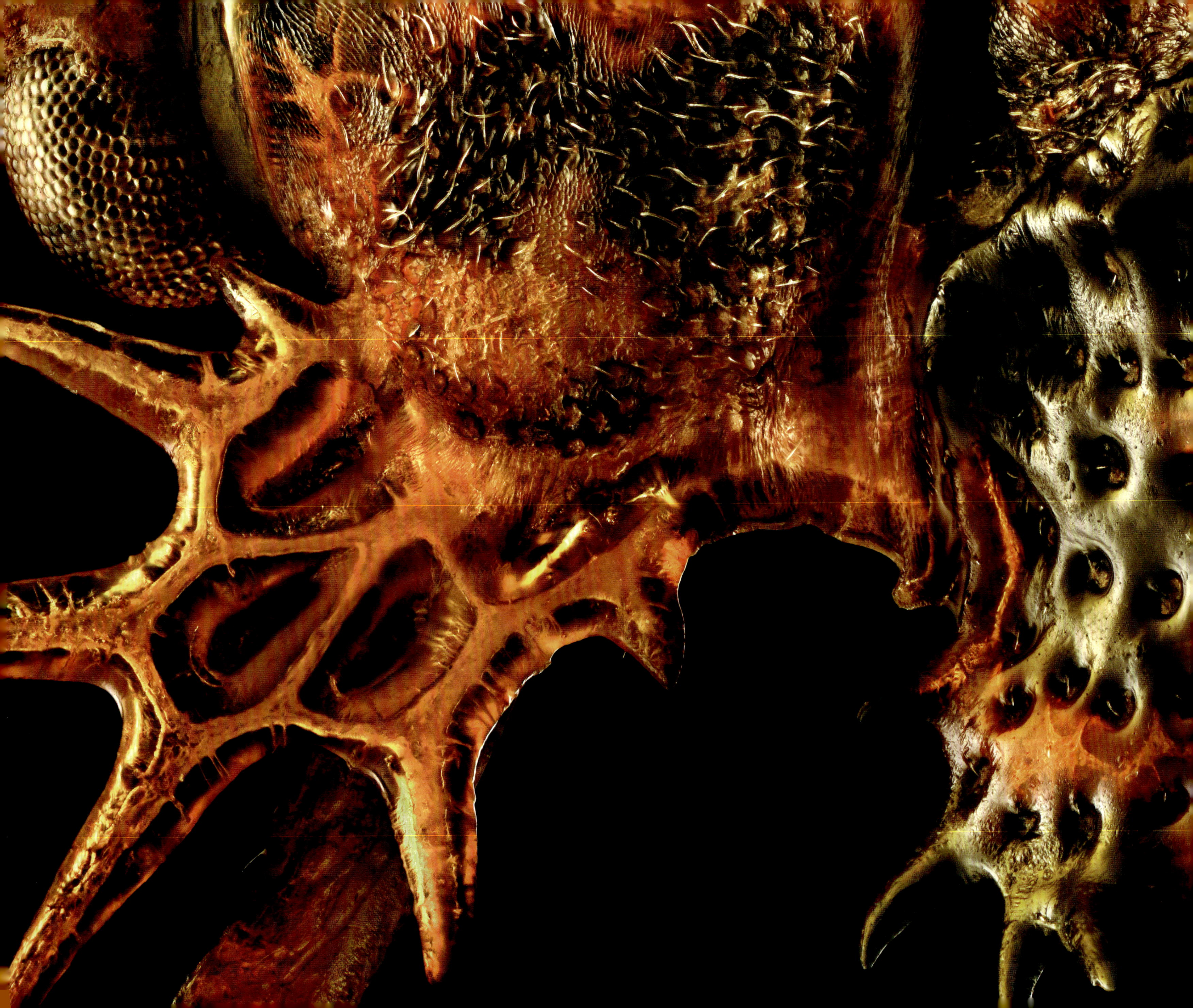

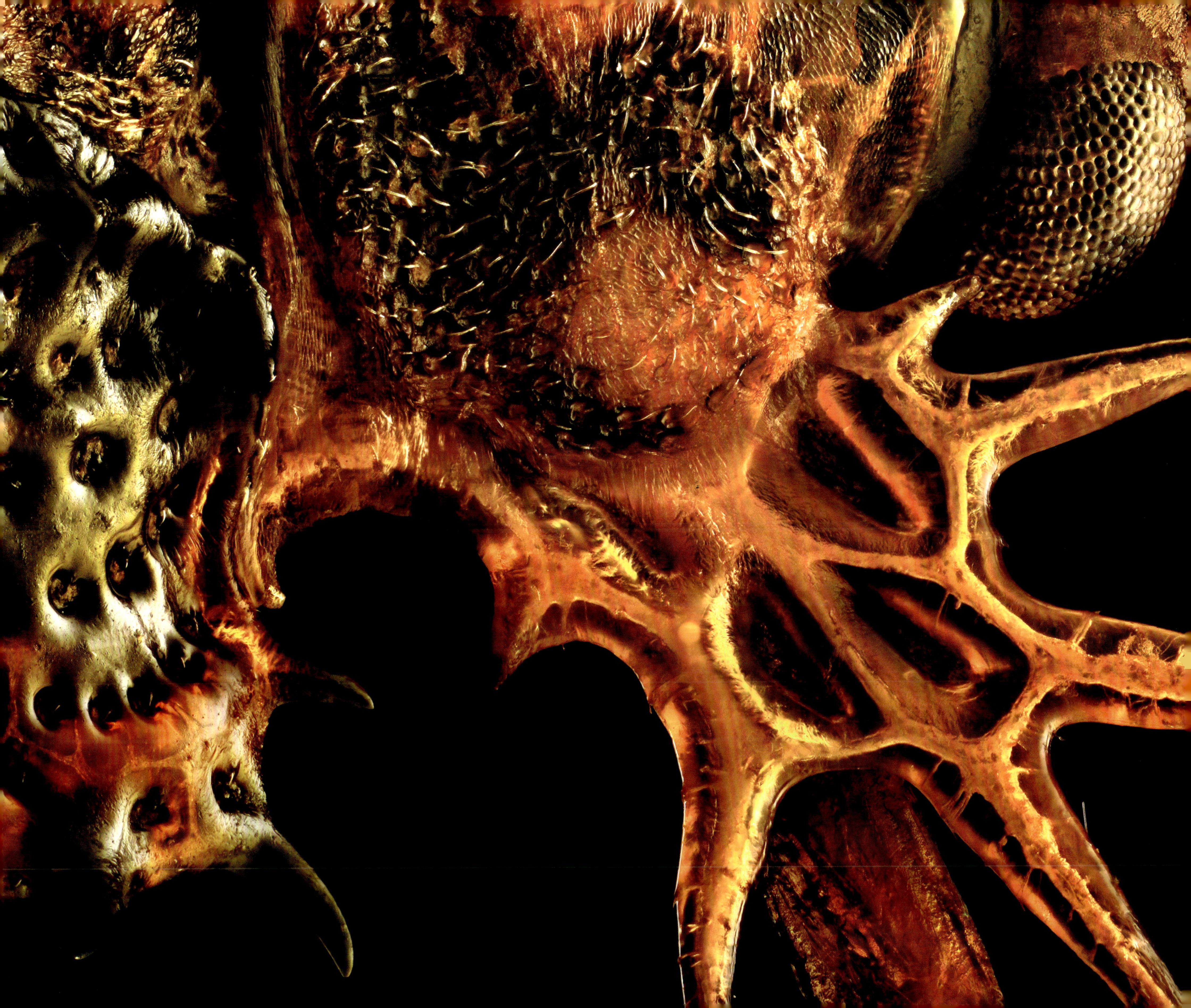

DARKLING BEETLE

Stenocara eburnea (Coleoptera, Tenebrionidae)

White coloration is relatively rare in beetles and its function in this species is uncertain. It could be an adaptation to reflect the strong sunlight of its desert habitat, preventing overheating. Many desert-inhabiting darkling beetles are able to spray noxious defensive chemicals from specialized abdominal glands, so it is possible the striking contrast between black and white serves as a warning pattern to predators.

NAMIBIA

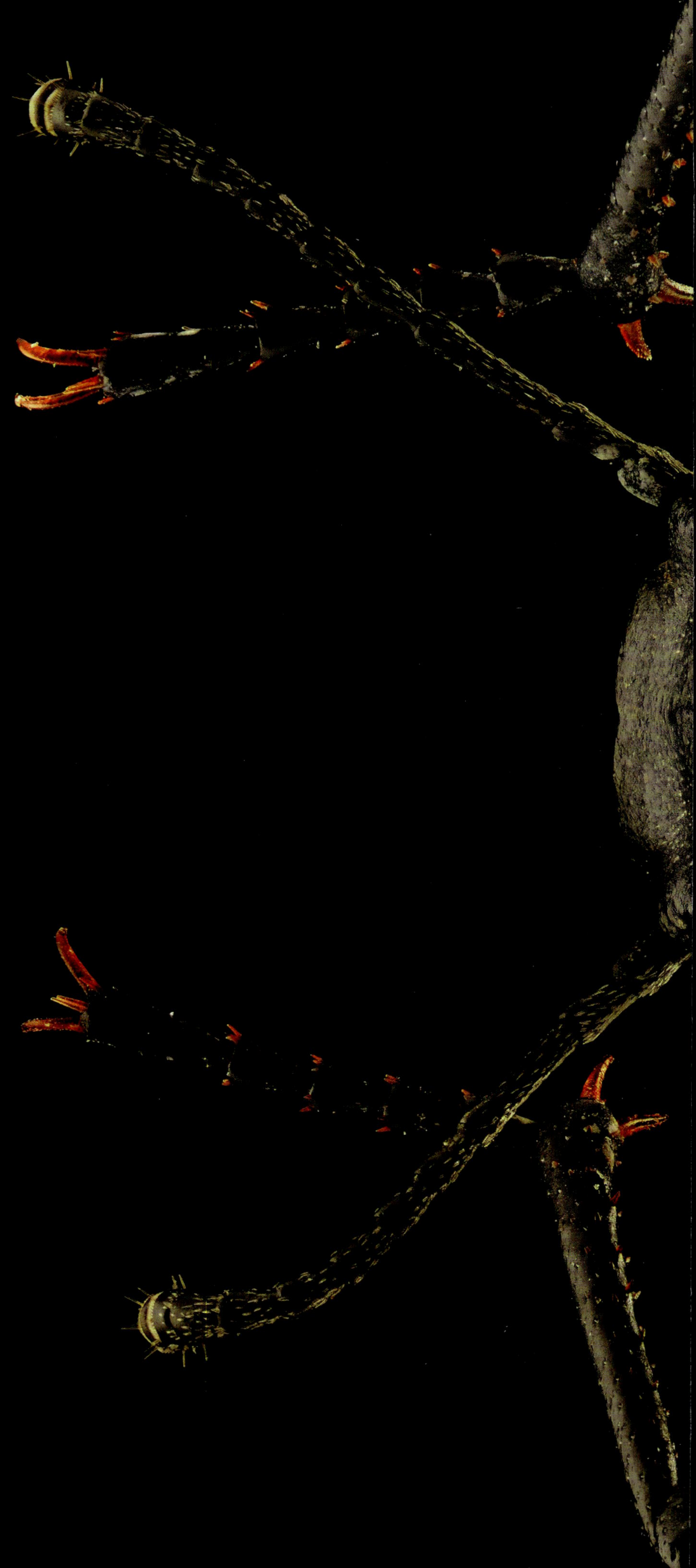

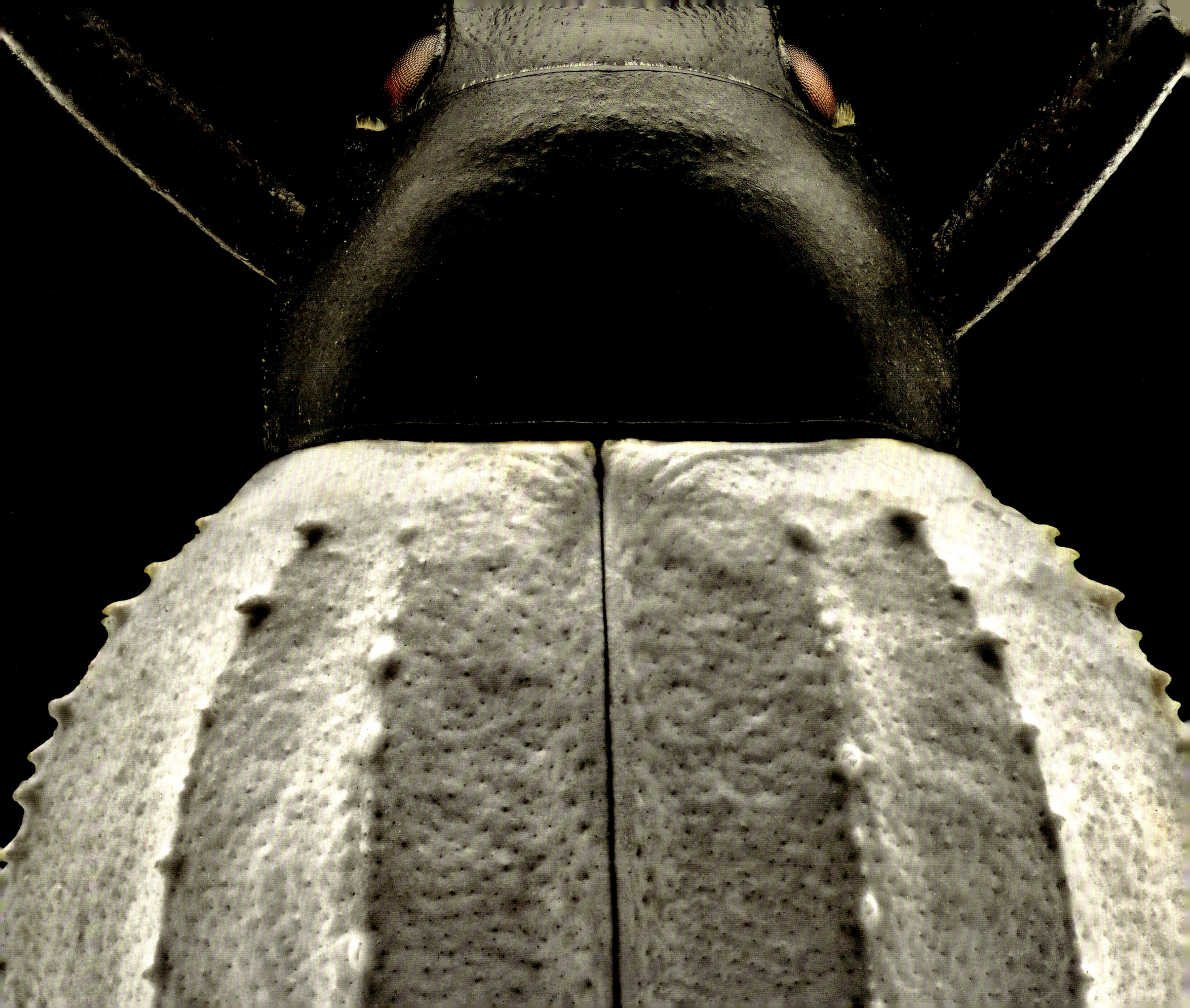

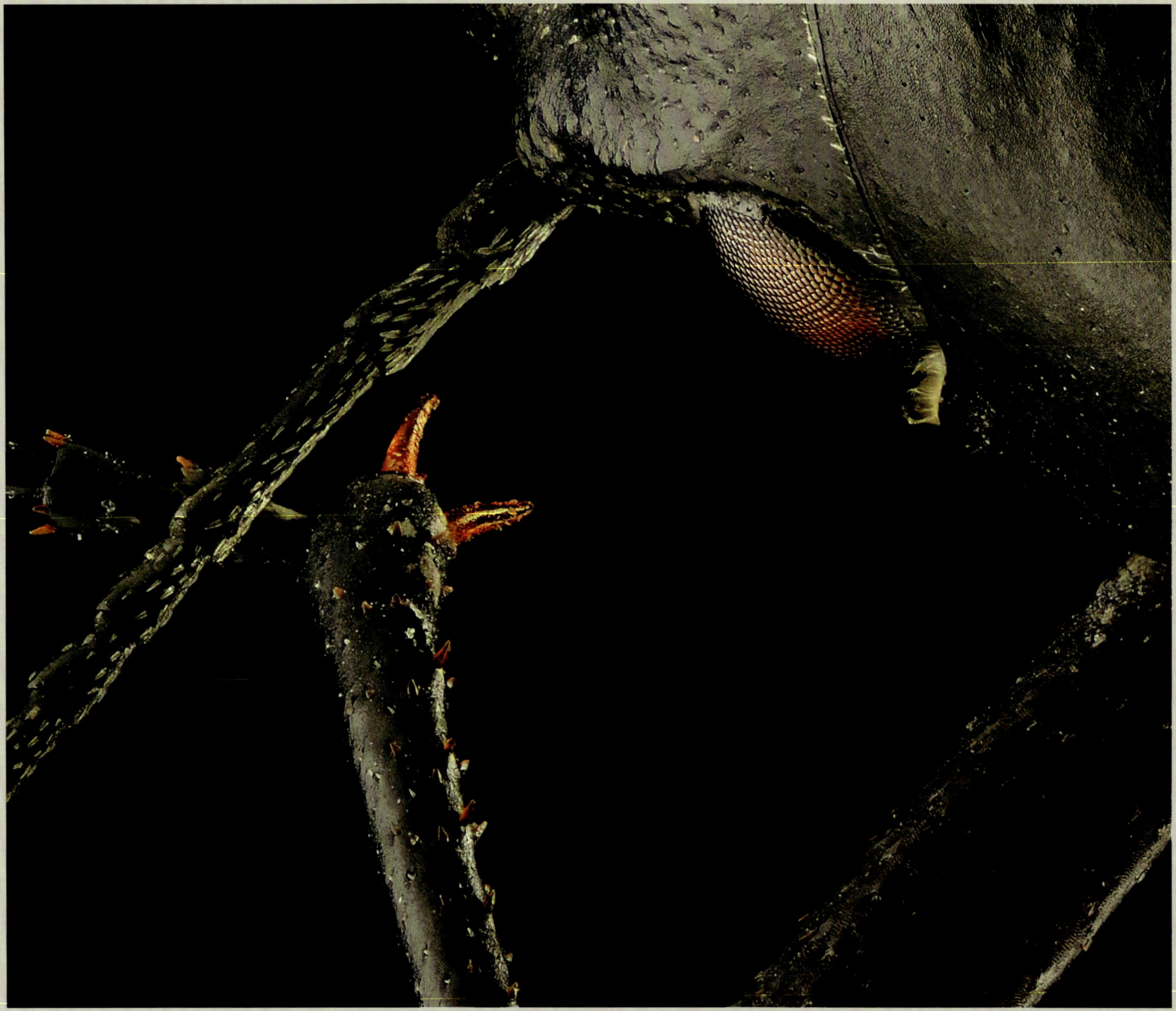

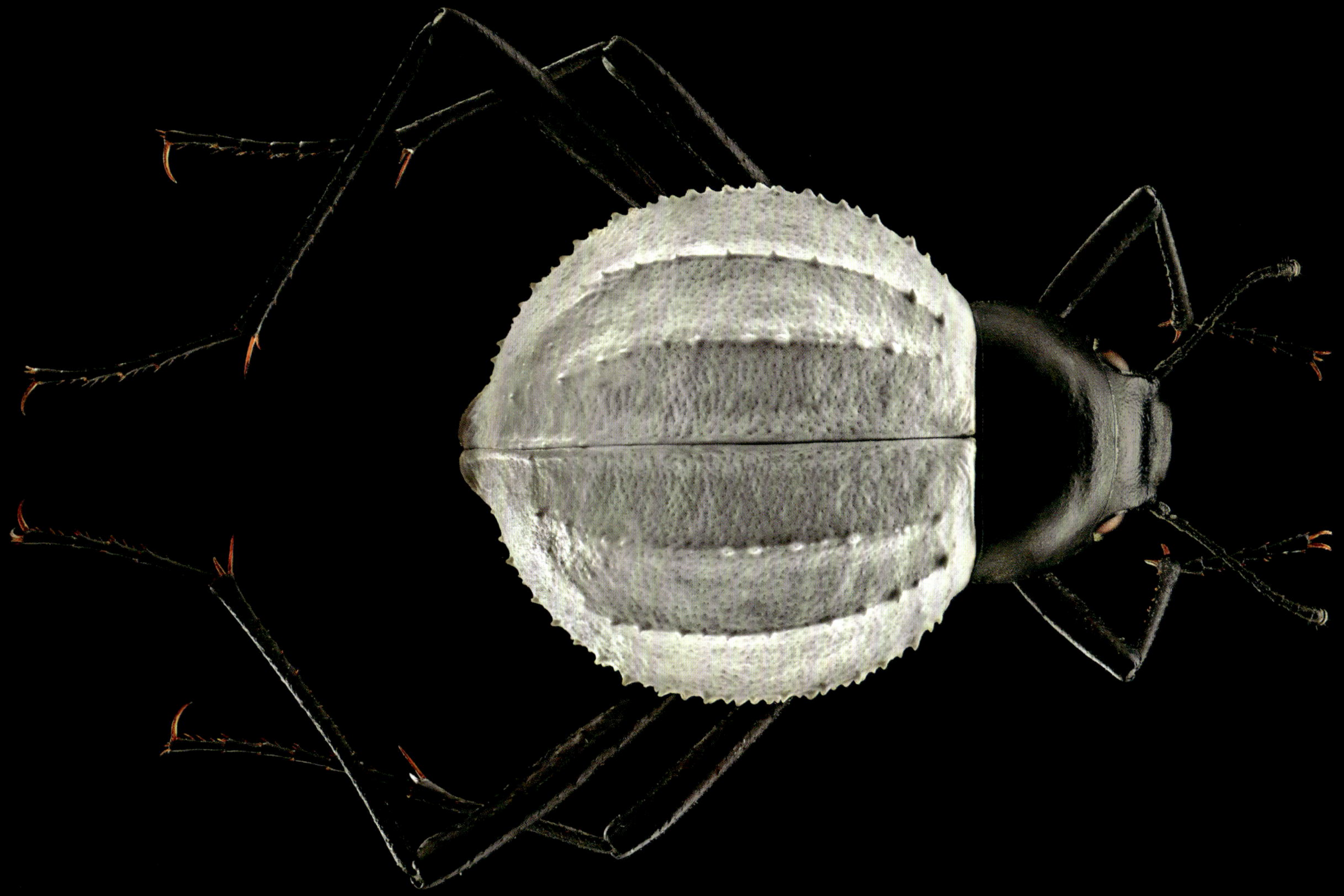

FLYING SAUCER TRENCH BEETLE

Lepidochora porti (Coleoptera, Tenebrionidae)

Native to the Namib Desert of southern Africa, this insect has evolved a remarkable method of harvesting water from coastal fog. First it digs a series of trenches and ridges in a suitably orientated sand dune. As the structure stands slightly taller than the surrounding ground, water condenses on the ridges and can then be sucked up by the beetle. Although invisible to the naked eye, higher magnification reveals the beautiful covering of minute white scales. Similar scales in other insects have a range of uses but in this case their function is unknown. The scales could reduce friction when burrowing, aid in heat regulation, or reduce water loss.

NAMIBIA

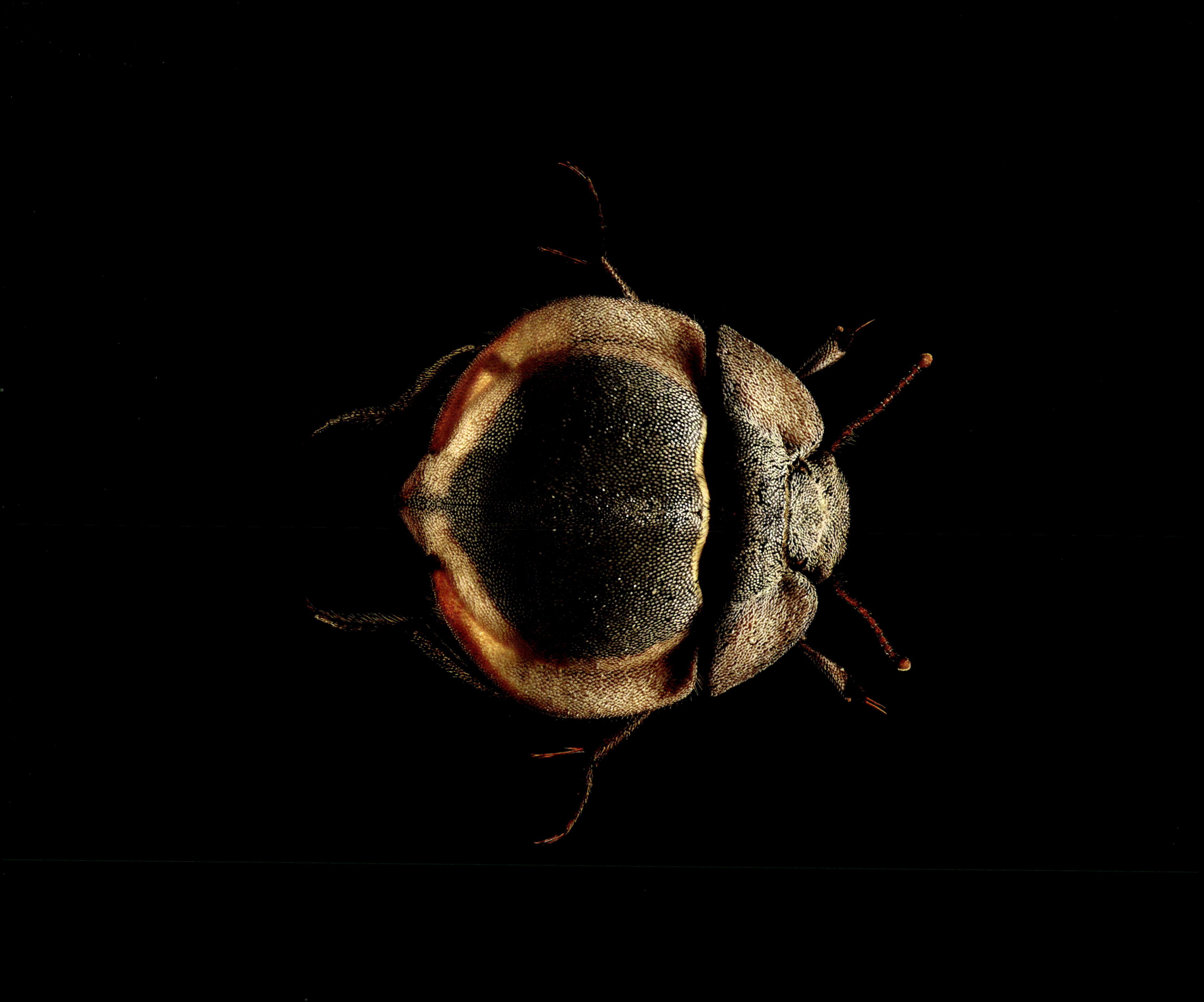

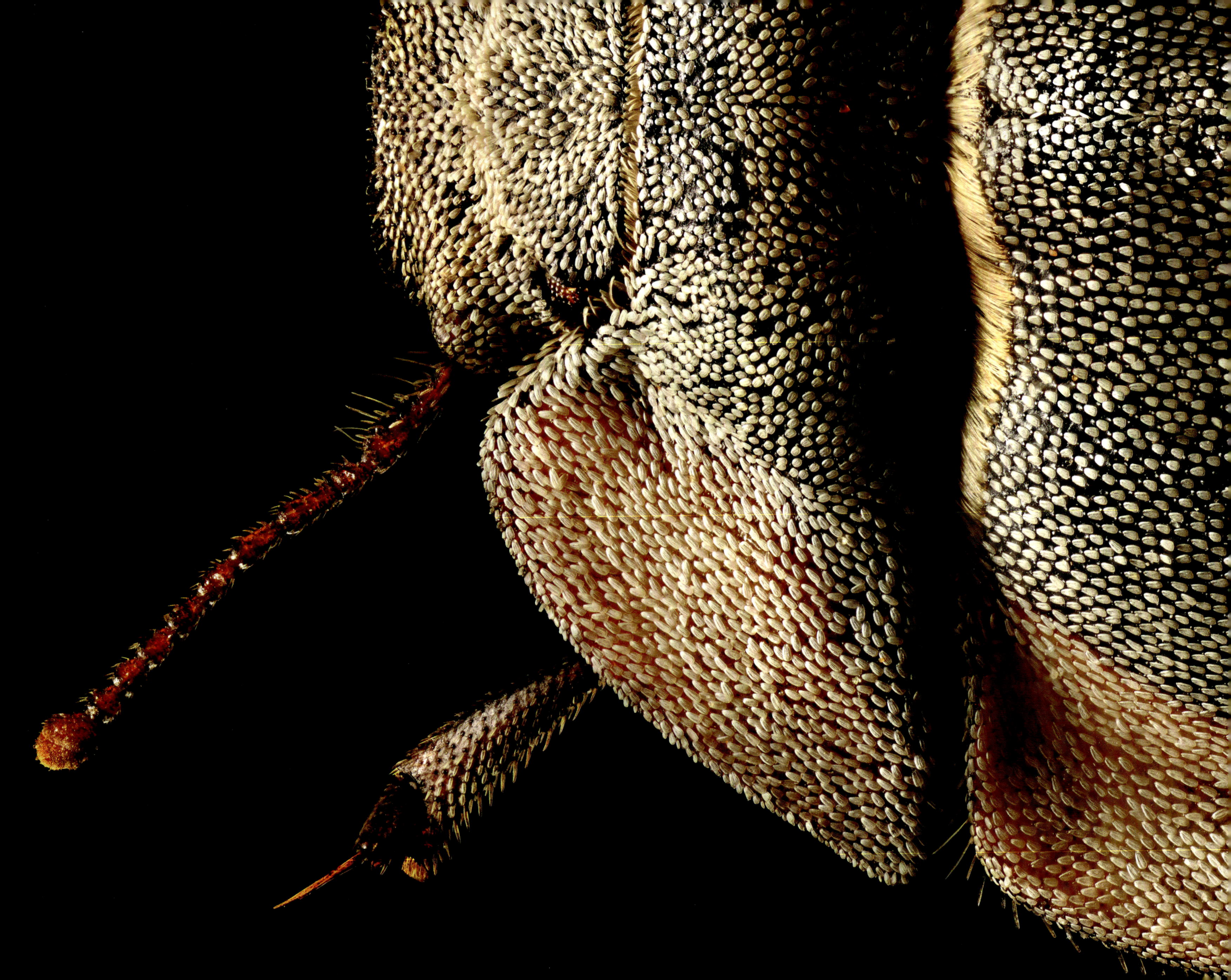

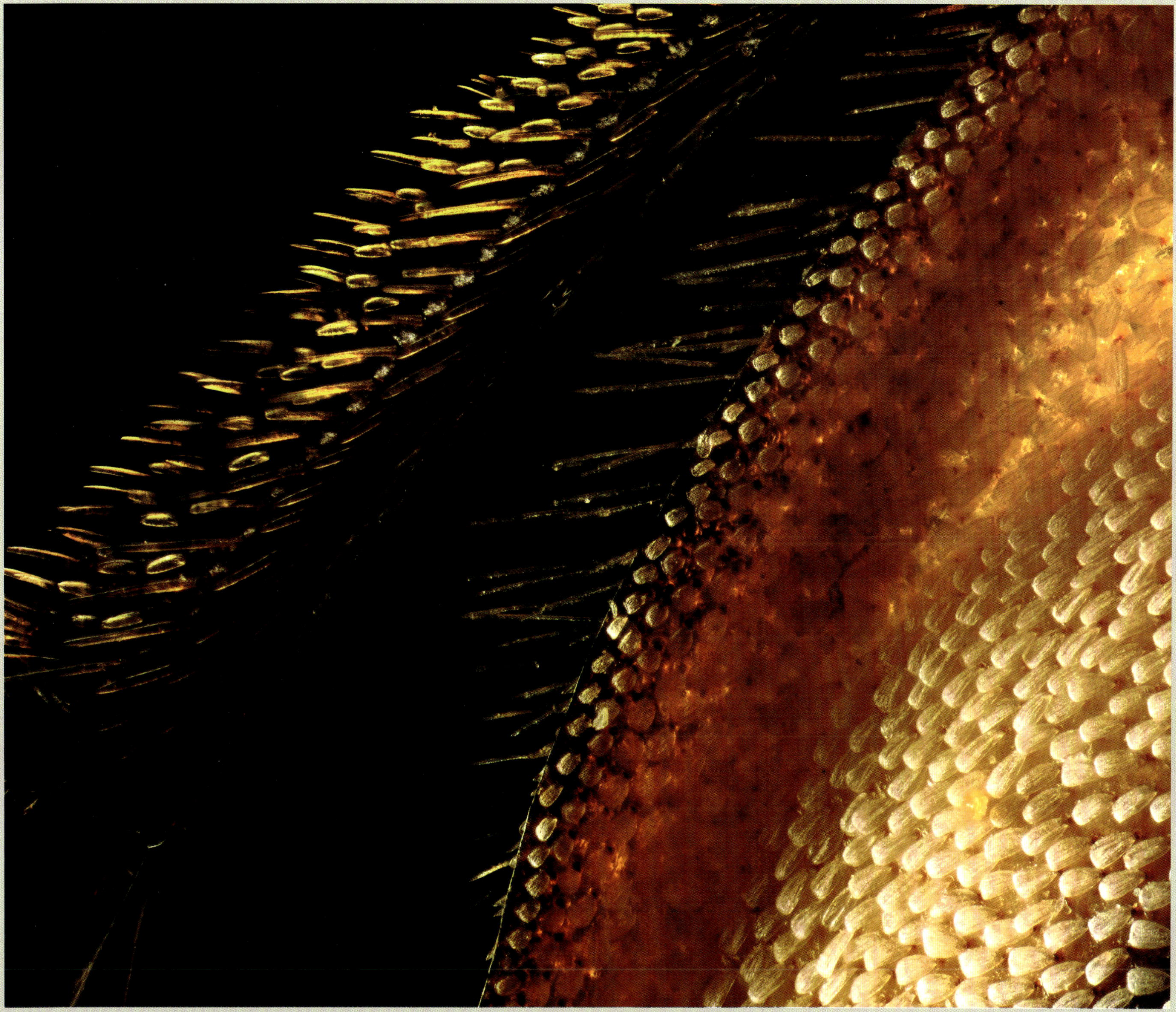

ORCHID CUCKOO BEE

Exaerete frontalis (Hymenoptera, Apidae)

The orchid cuckoo bee is one of the most spectacular of all bees in terms of size, color, and microsculpture. We usually think of bees as benign, helpful creatures, but *Exaerete* is a cuckoo bee. Instead of collecting pollen and constructing their own nests, female cuckoo bees enter the nests of other bees to lay their own eggs in the host's brood cells. This strategy is so successful that a significant proportion of all bee species are parasites in this way. Not having to collect pollen, cuckoo bees are often less hairy than other bees and generally have a more wasp-like appearance.

BRAZIL

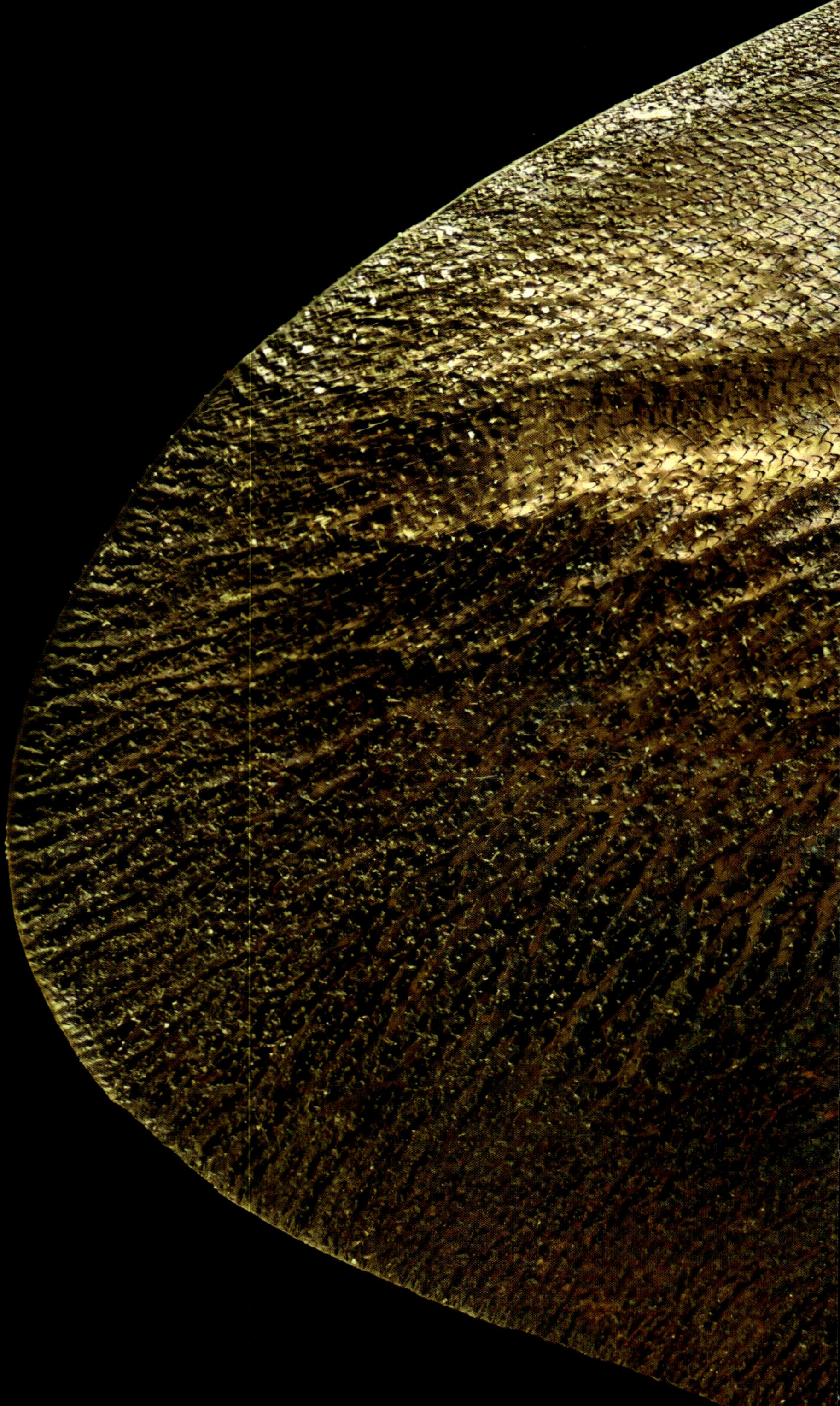

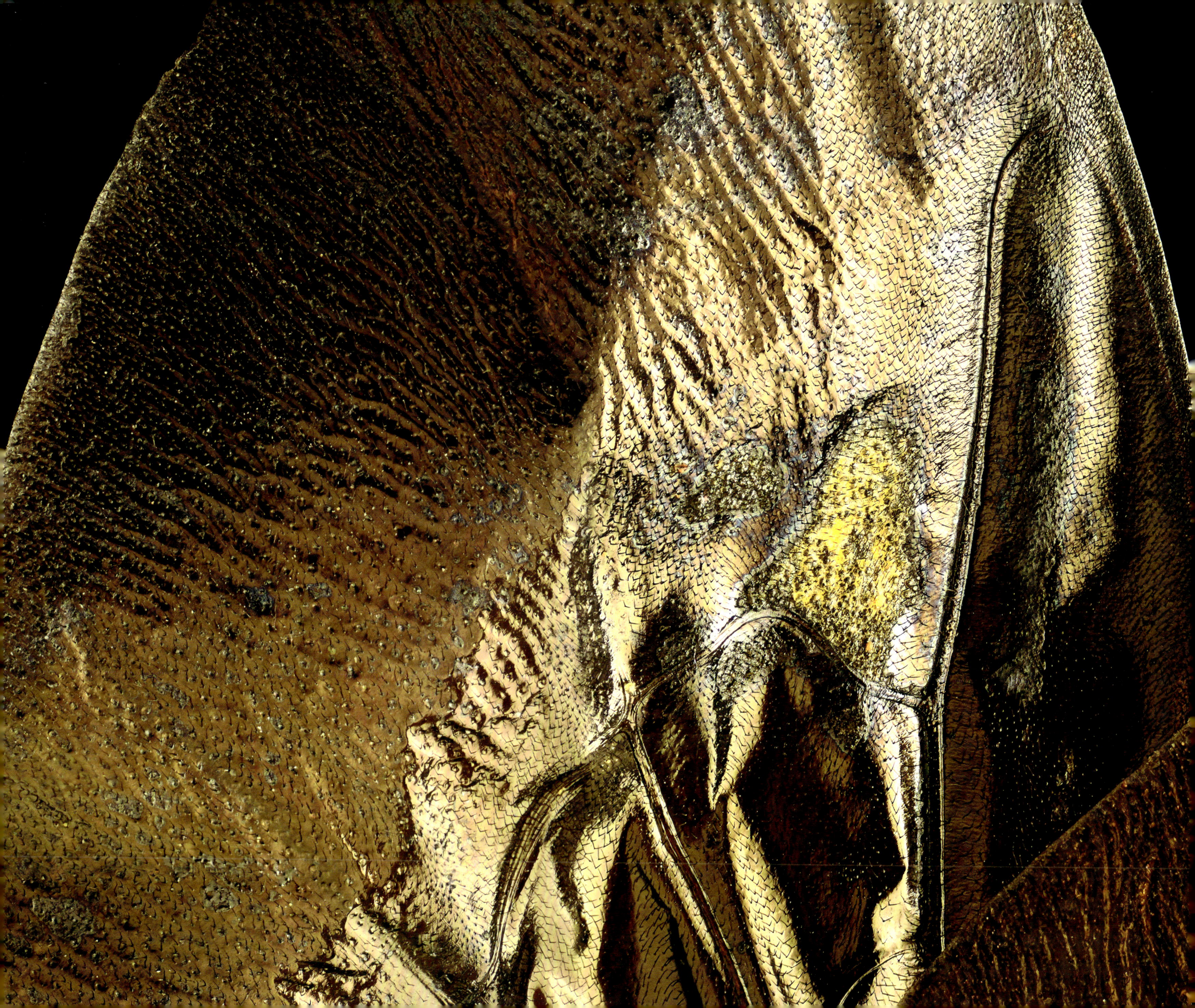

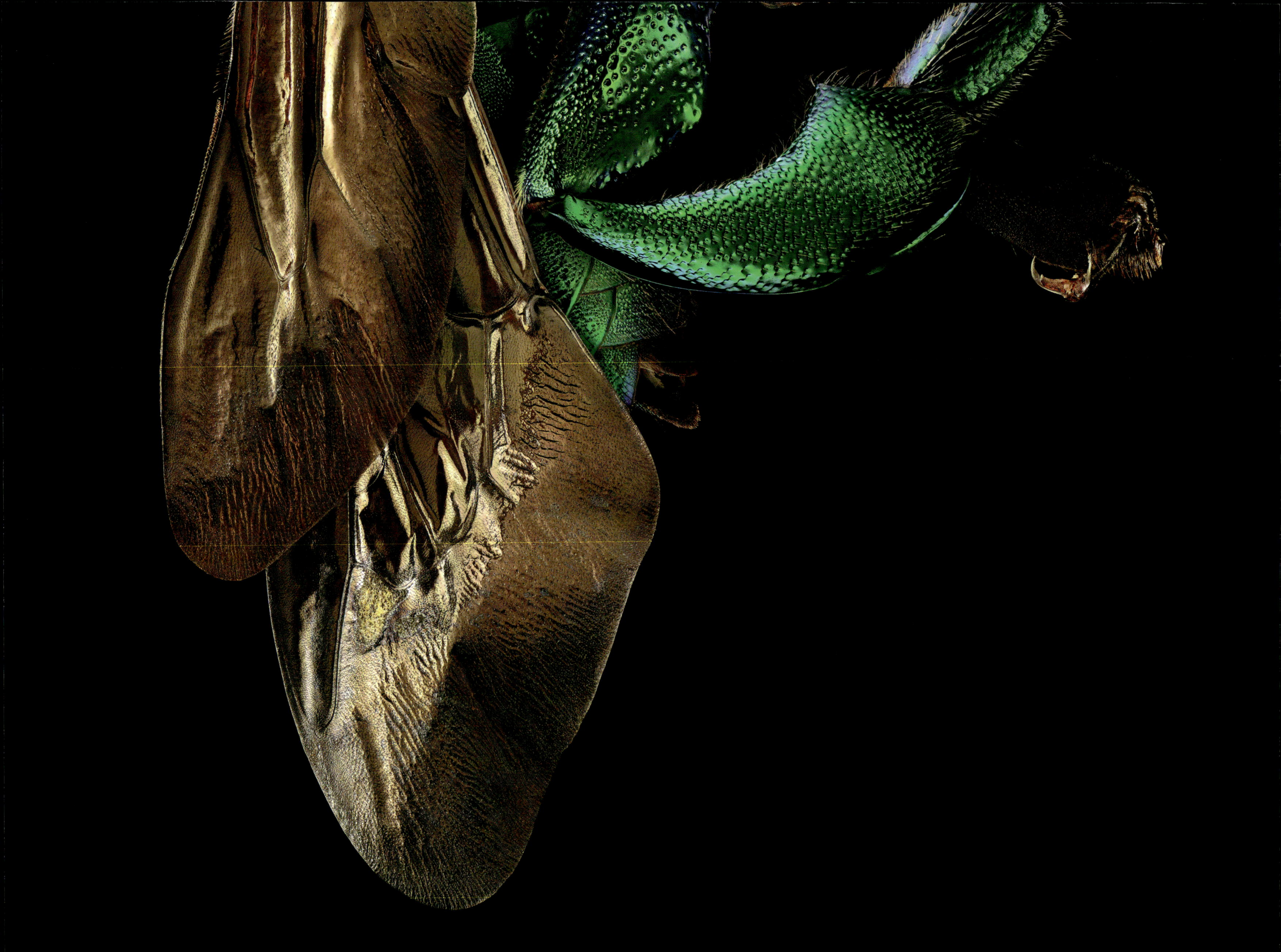

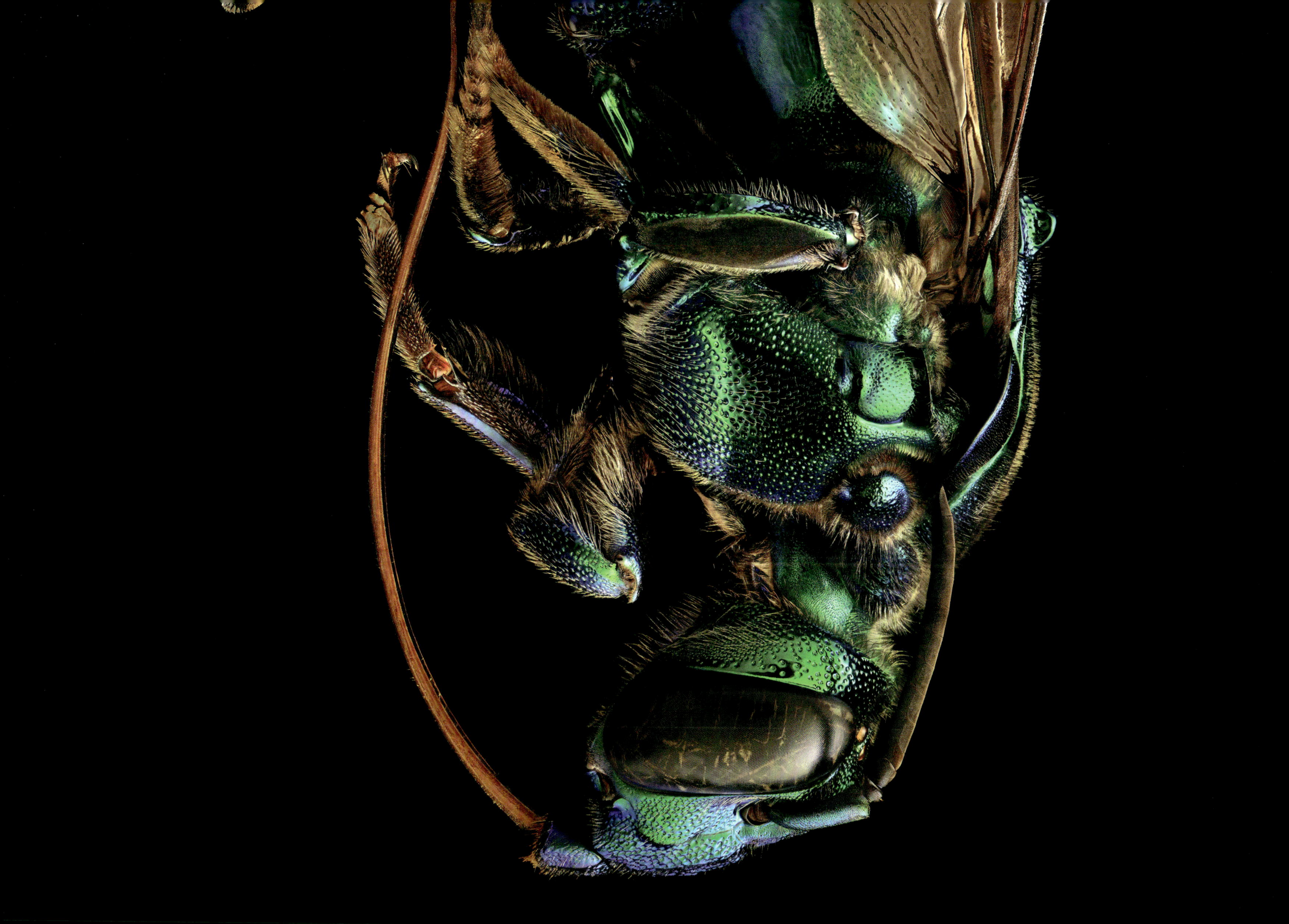

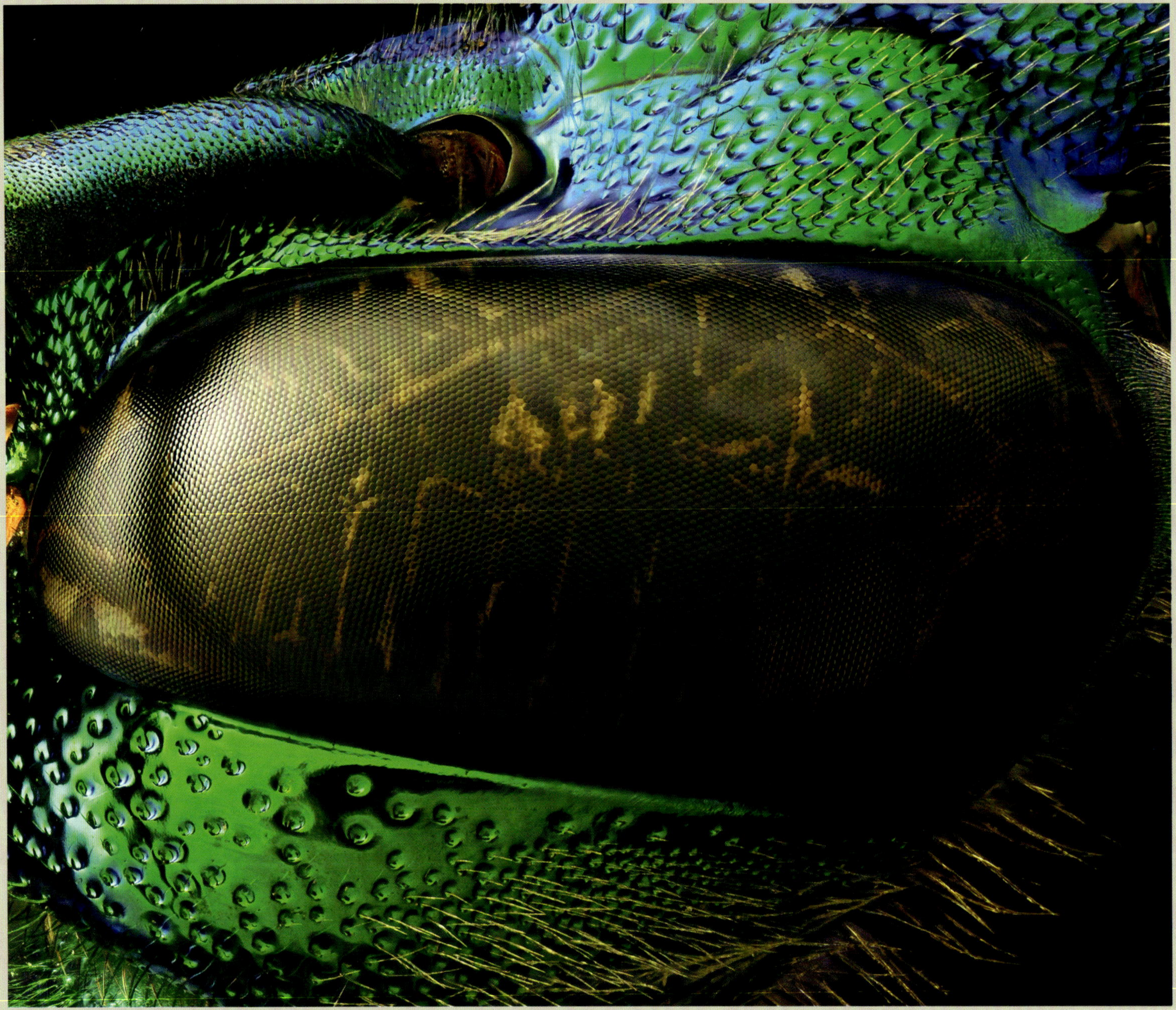

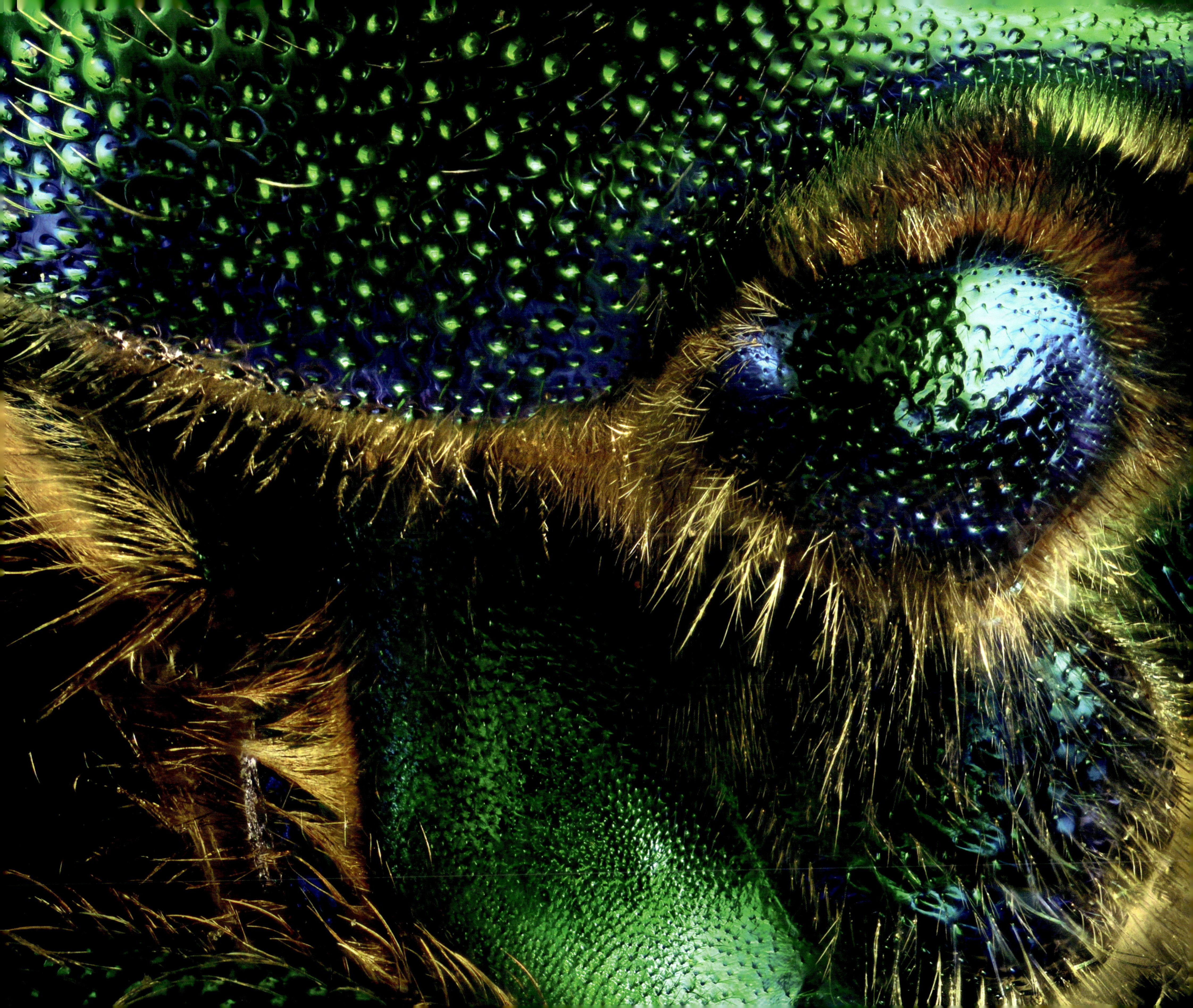

TREEHOPPER

SPECIES OF THE GENUS *Phyllotropis*
(Hemiptera, Membracidae)

Treehoppers show a very wide diversity of shape and form, arguably more than in any other group of insects. The thorax is usually modified in some way to provide protection for the treehopper. It can be shaped like a thorn or leaf for camouflage and in a few species the thorax is shaped like an ant. The dome-like shield of this particular treehopper is effectively used to display its bright orange and black warning colors.

BRAZIL

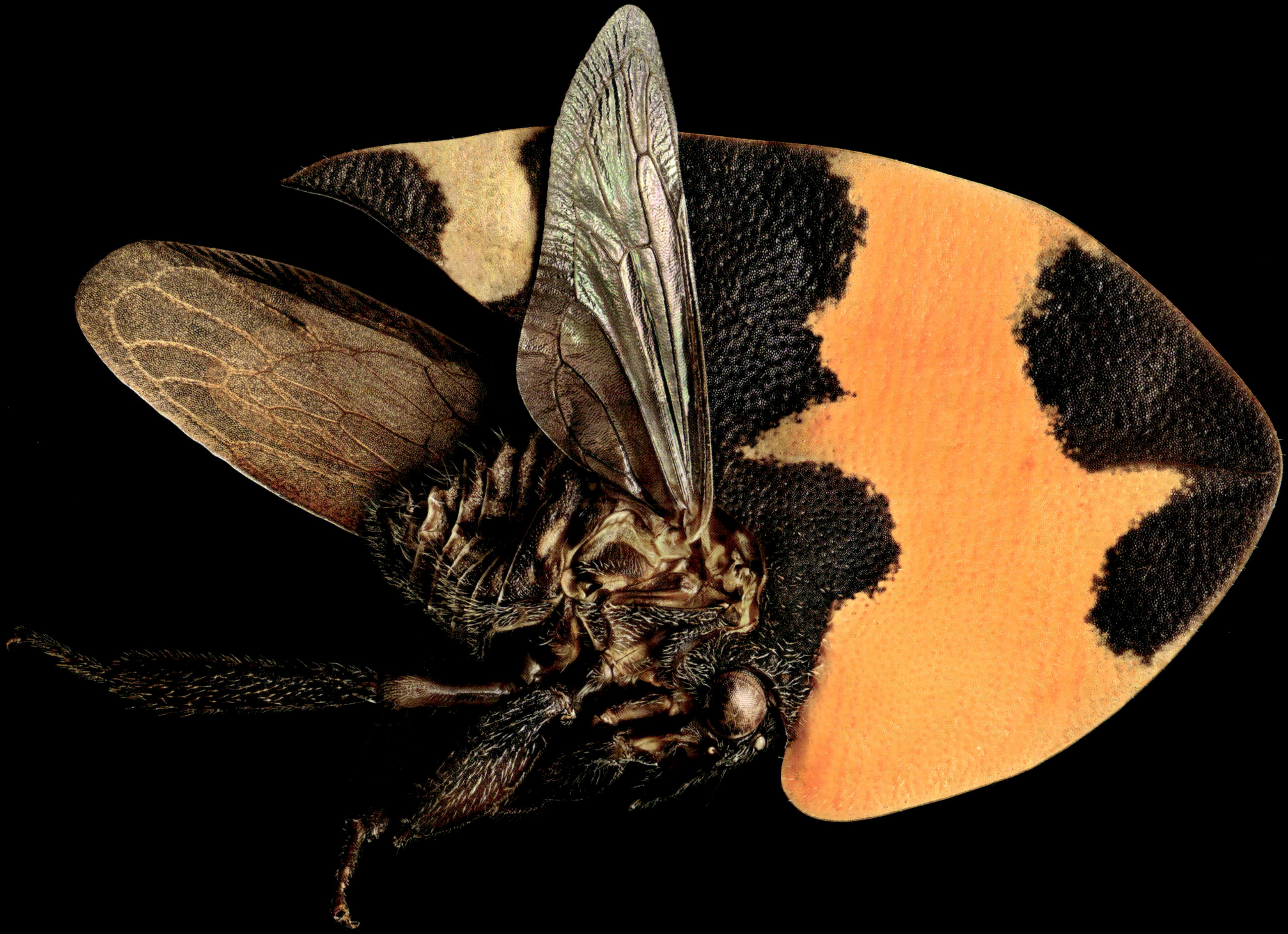

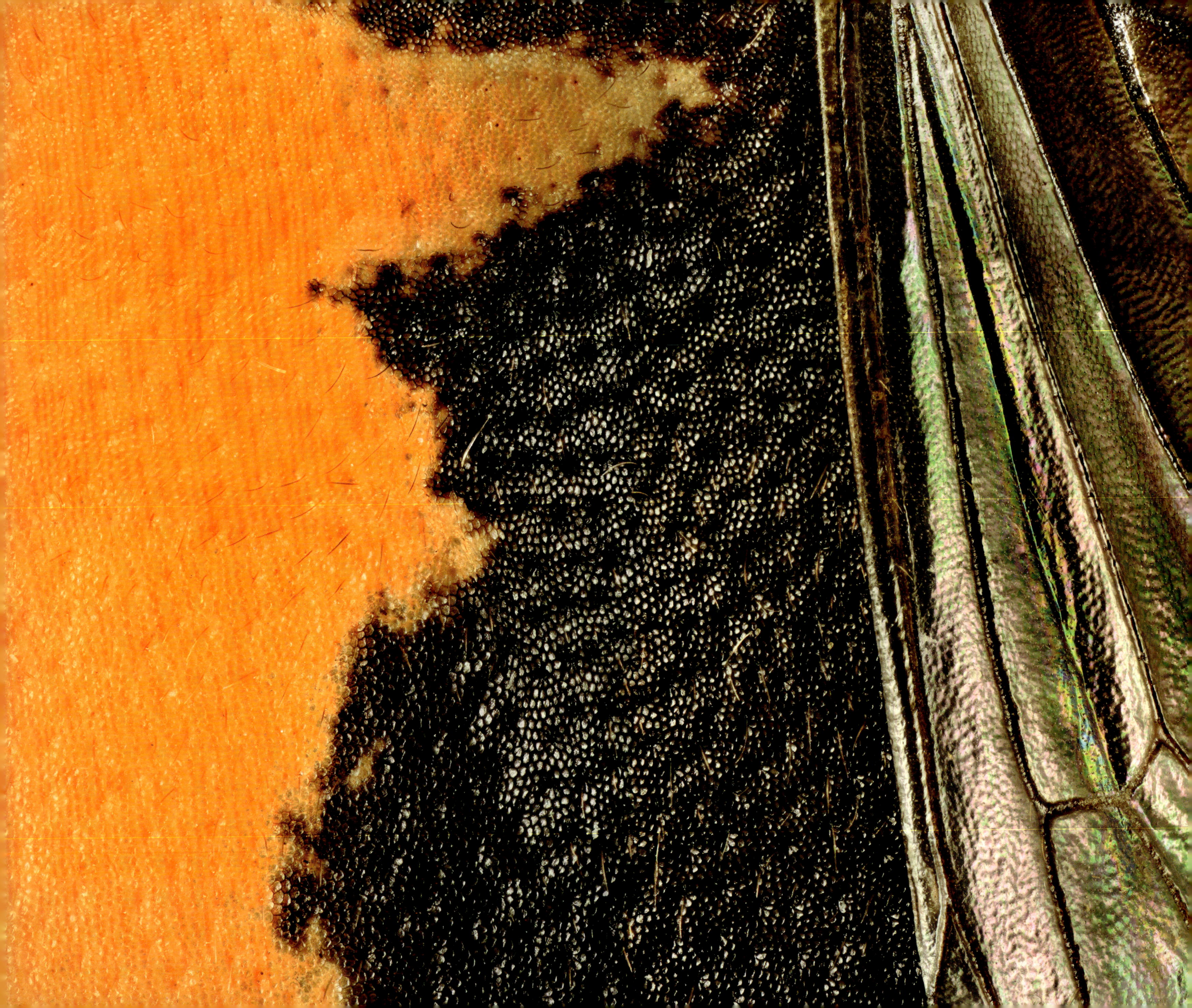

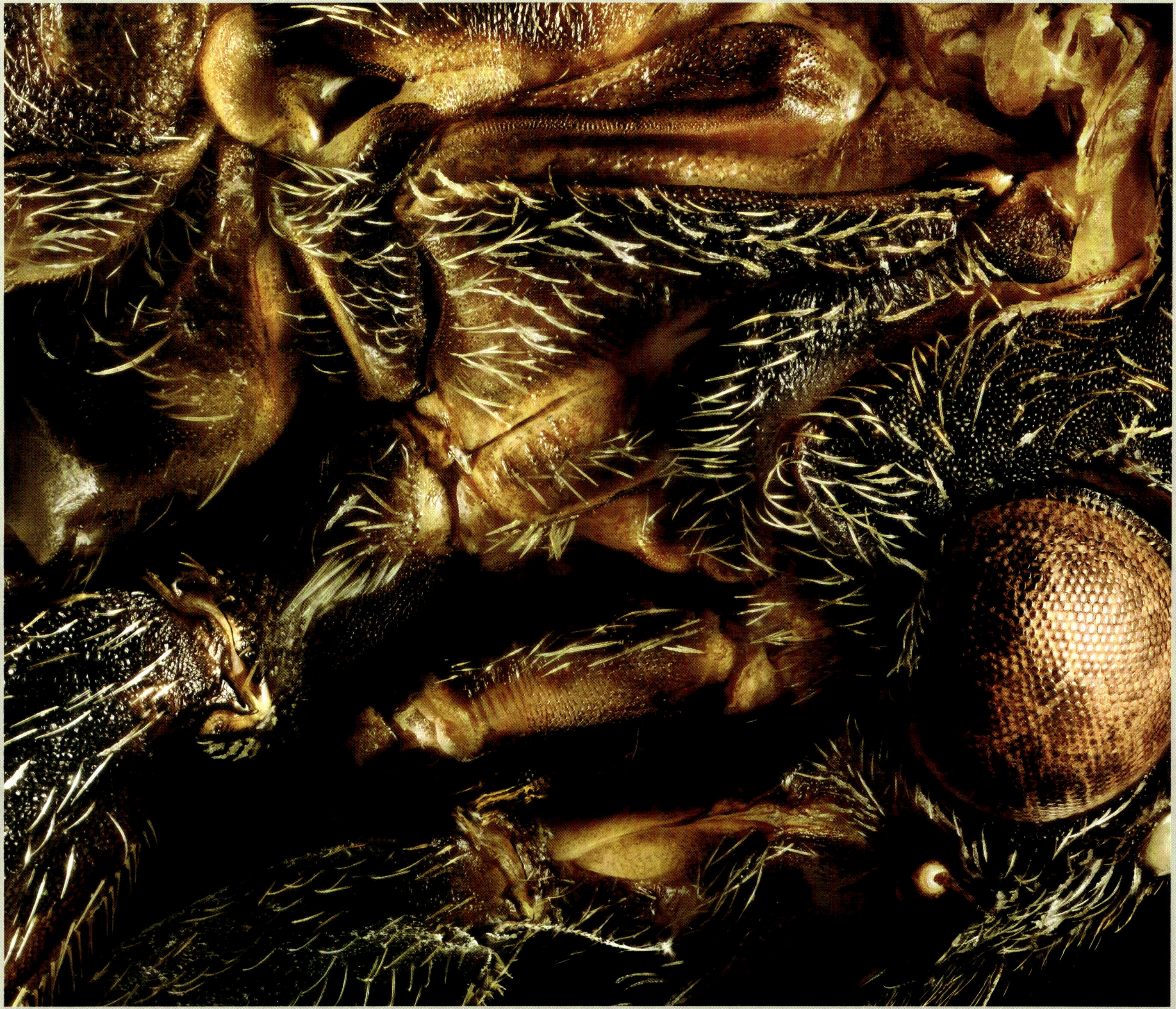

LONGHORN BEETLE

SPECIES OF THE GENUS *Sternotomis*
(Coleoptera, Cerambycidae)

Longhorns are another of the mega-diverse beetle groups, with more than twenty-five thousand species, showing a seemingly endless variety of shapes, sizes, and colors. Longhorn beetle larvae usually consume dead wood, a nutritionally poor substance. This can lead to exceptionally long development times for the larvae, sometimes in excess of twenty years. Once magnified, the secret to the spectacular patterning of this African longhorn is revealed—a covering of extremely fine pigmented scales similar to those of butterflies and moths.

NIGERIA

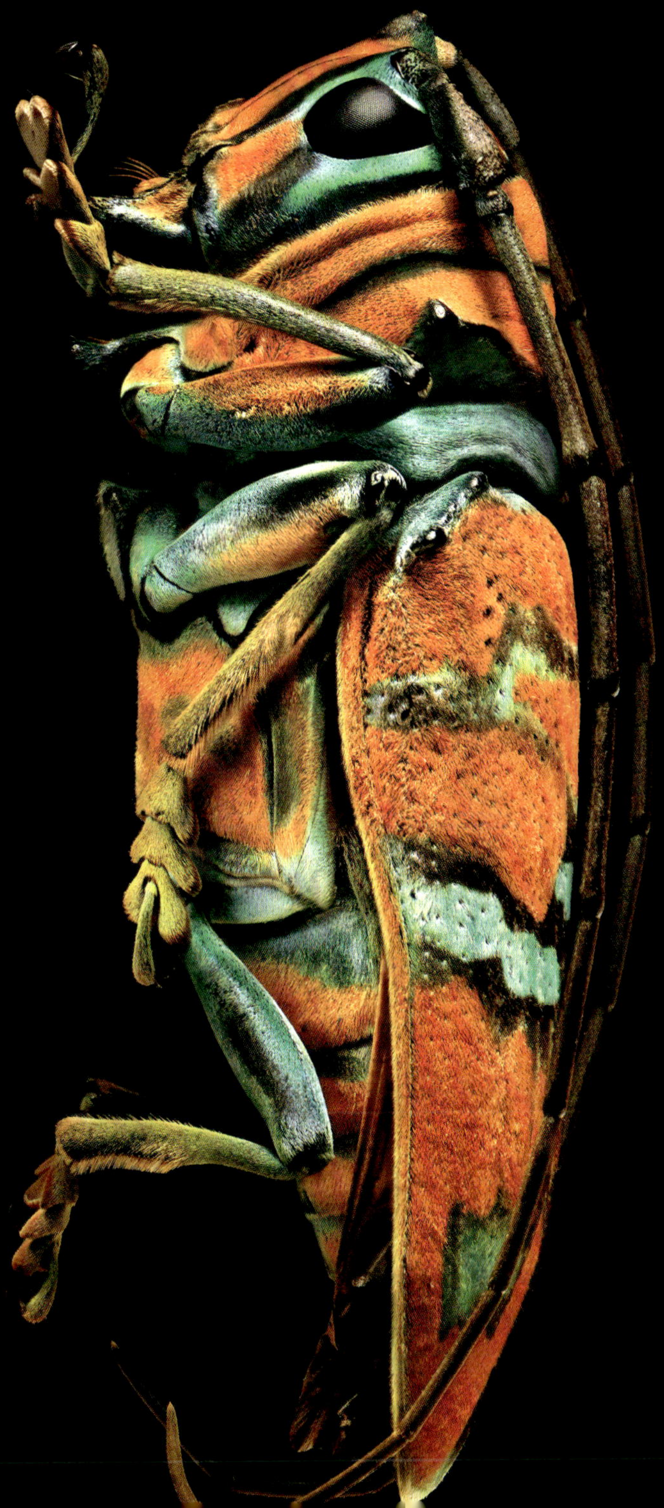

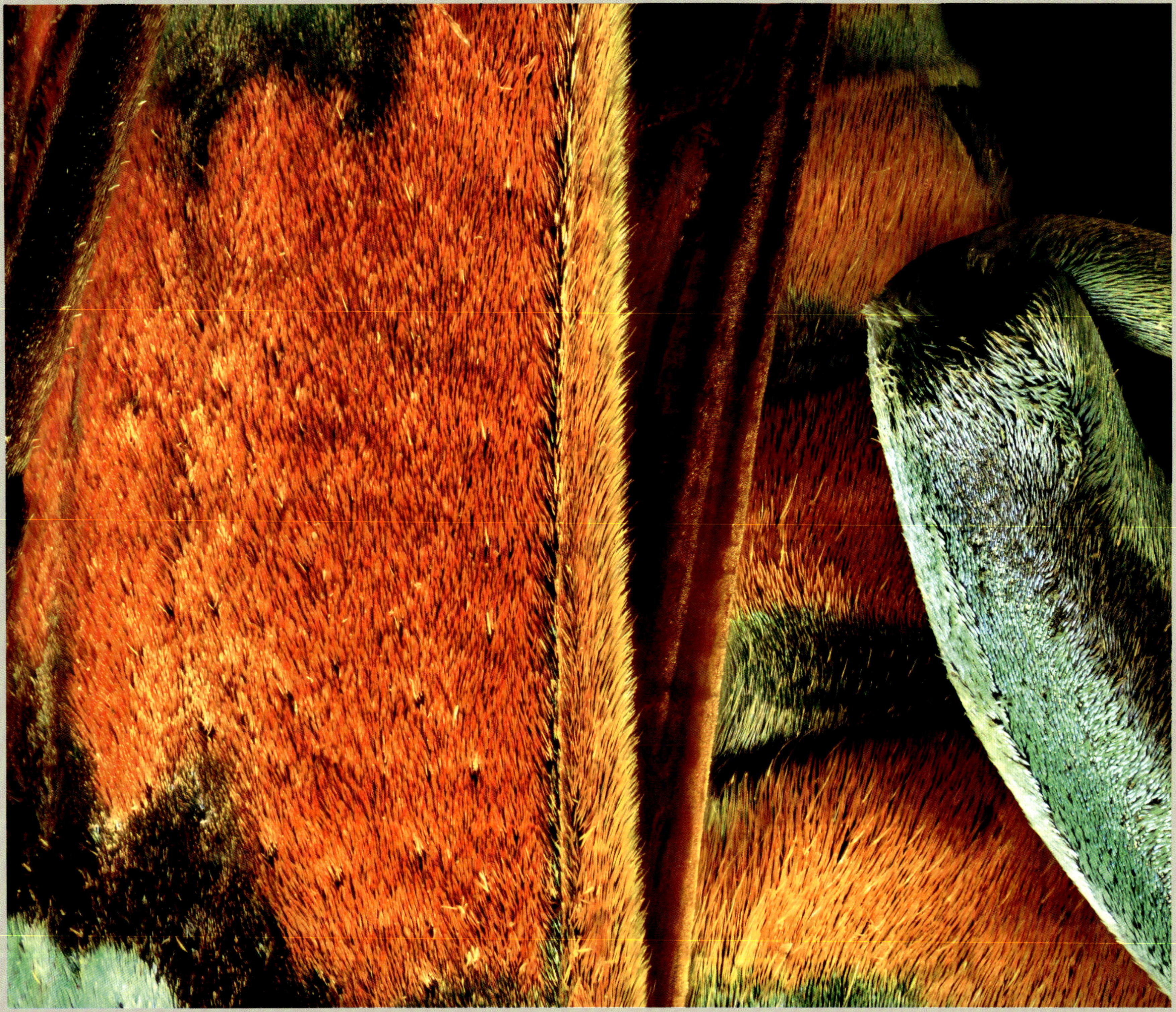

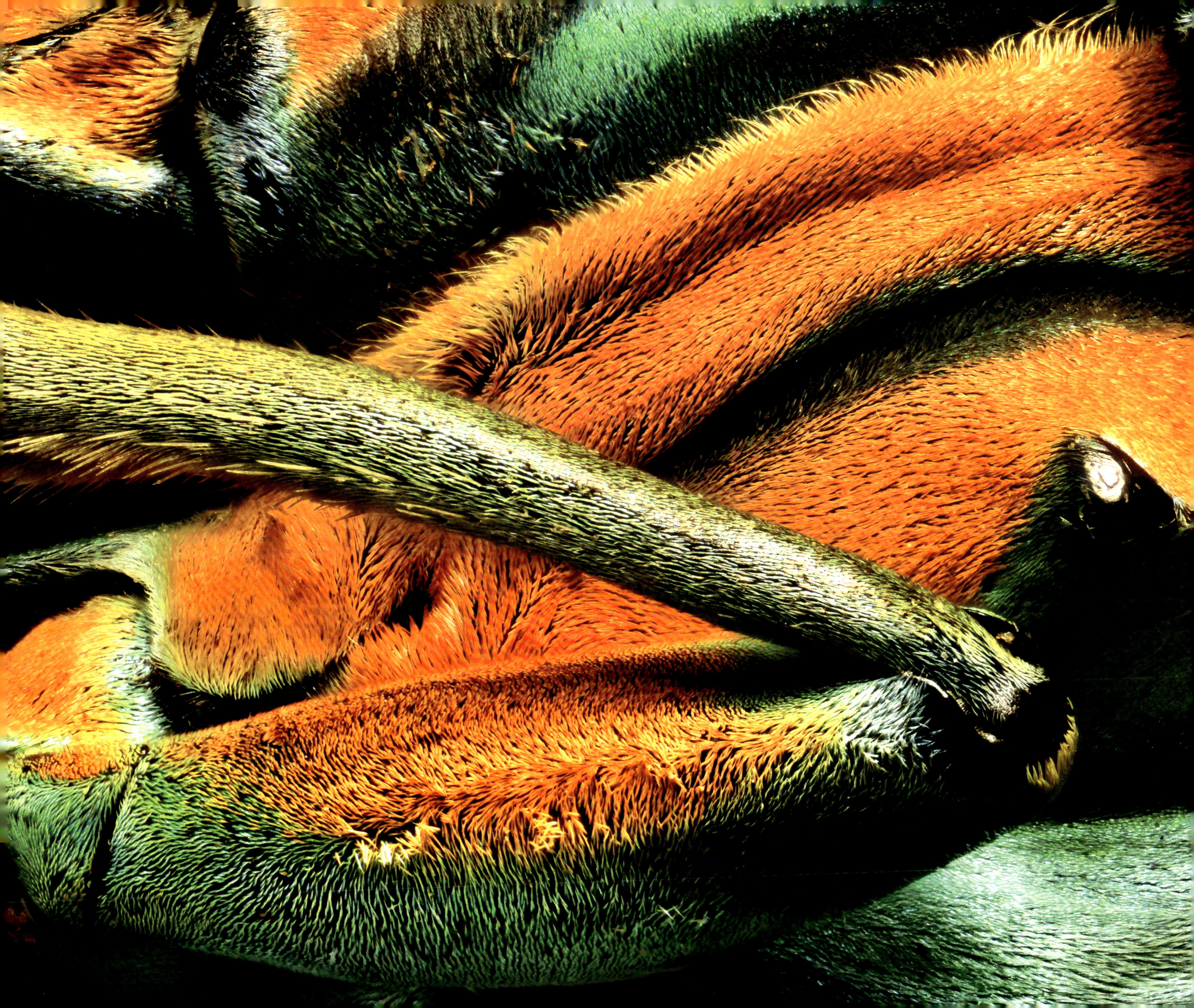

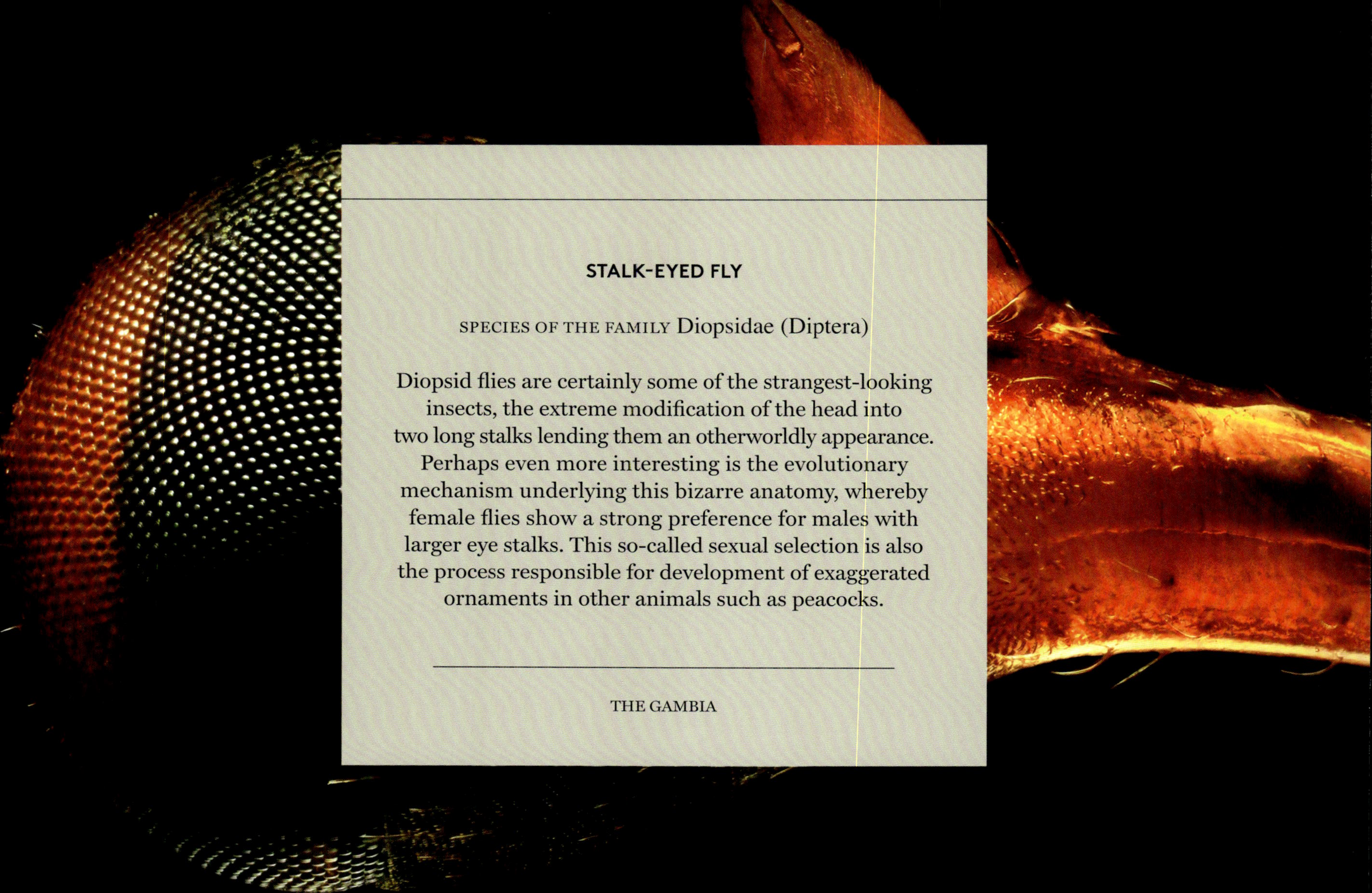

STALK-EYED FLY

SPECIES OF THE FAMILY Diopsidae (Diptera)

Diopsid flies are certainly some of the strangest-looking insects, the extreme modification of the head into two long stalks lending them an otherworldly appearance. Perhaps even more interesting is the evolutionary mechanism underlying this bizarre anatomy, whereby female flies show a strong preference for males with larger eye stalks. This so-called sexual selection is also the process responsible for development of exaggerated ornaments in other animals such as peacocks.

THE GAMBIA

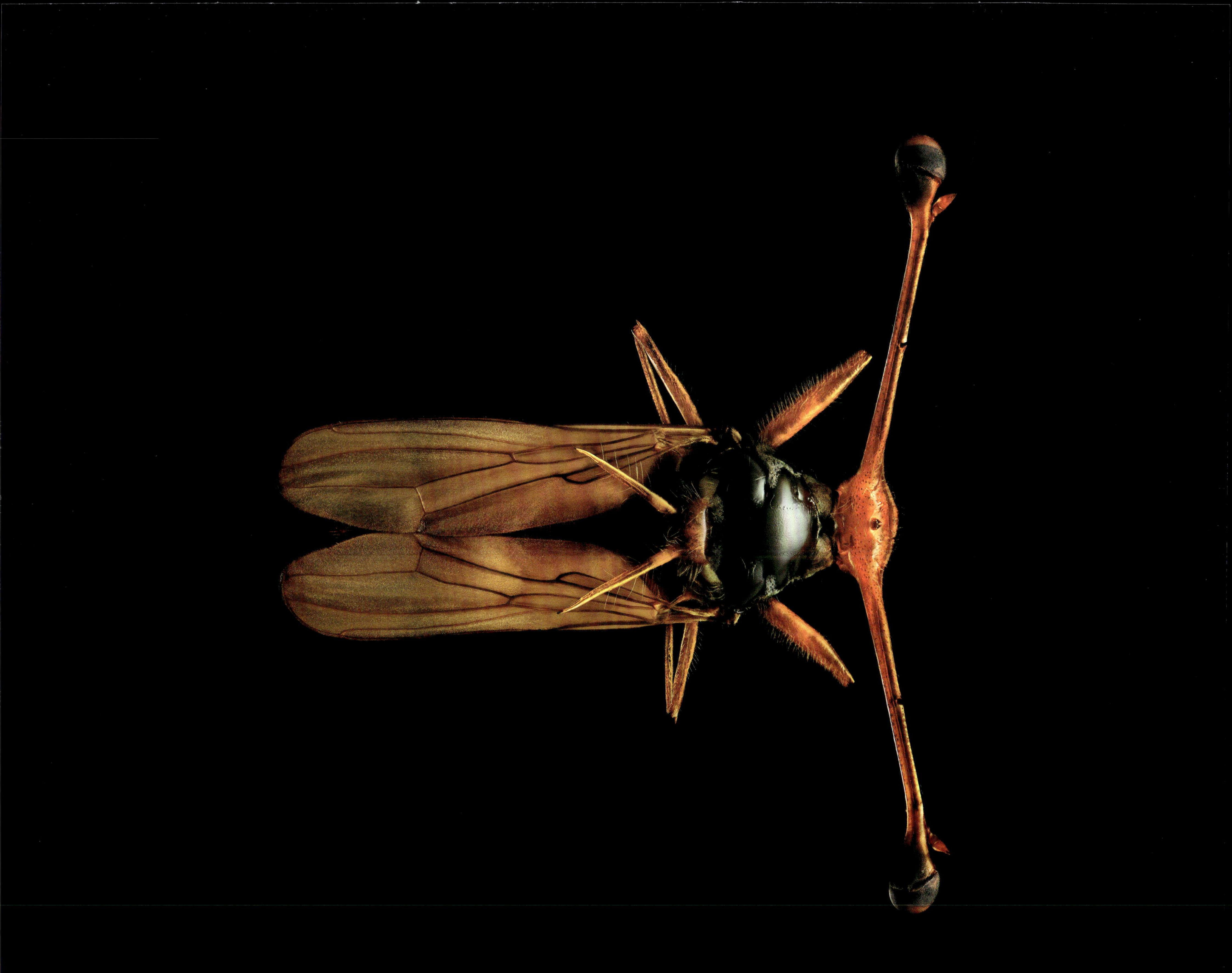

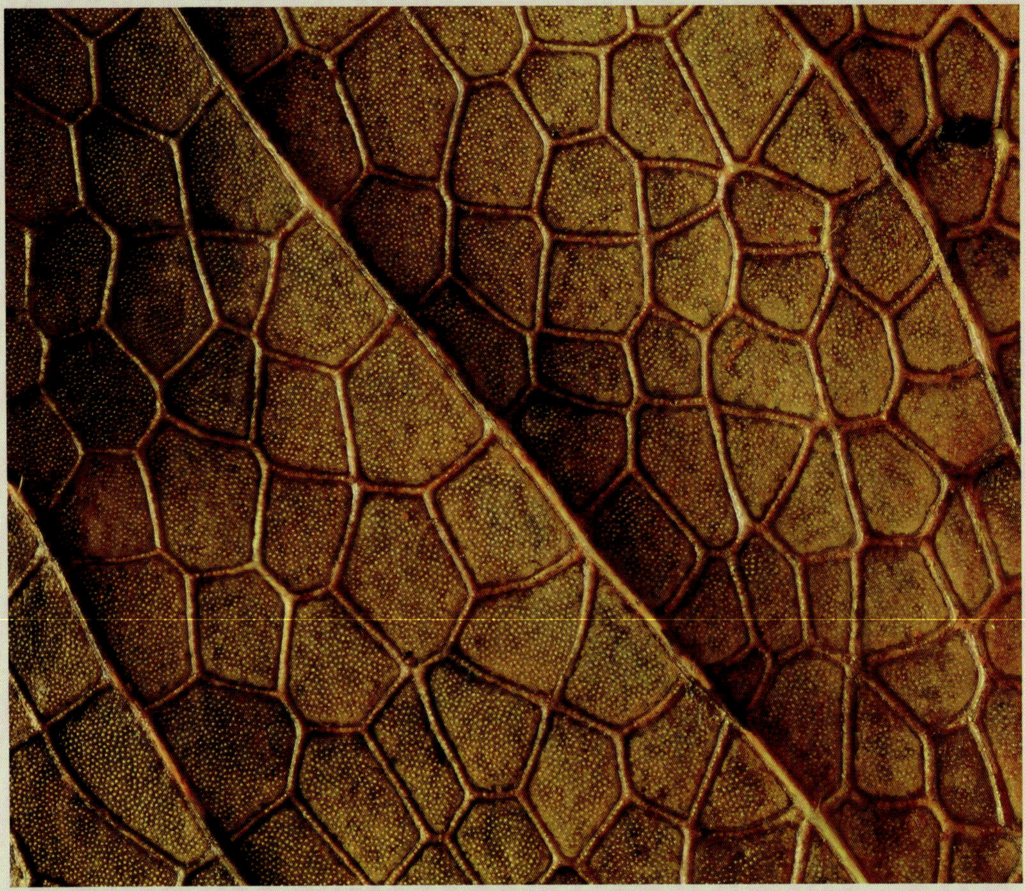
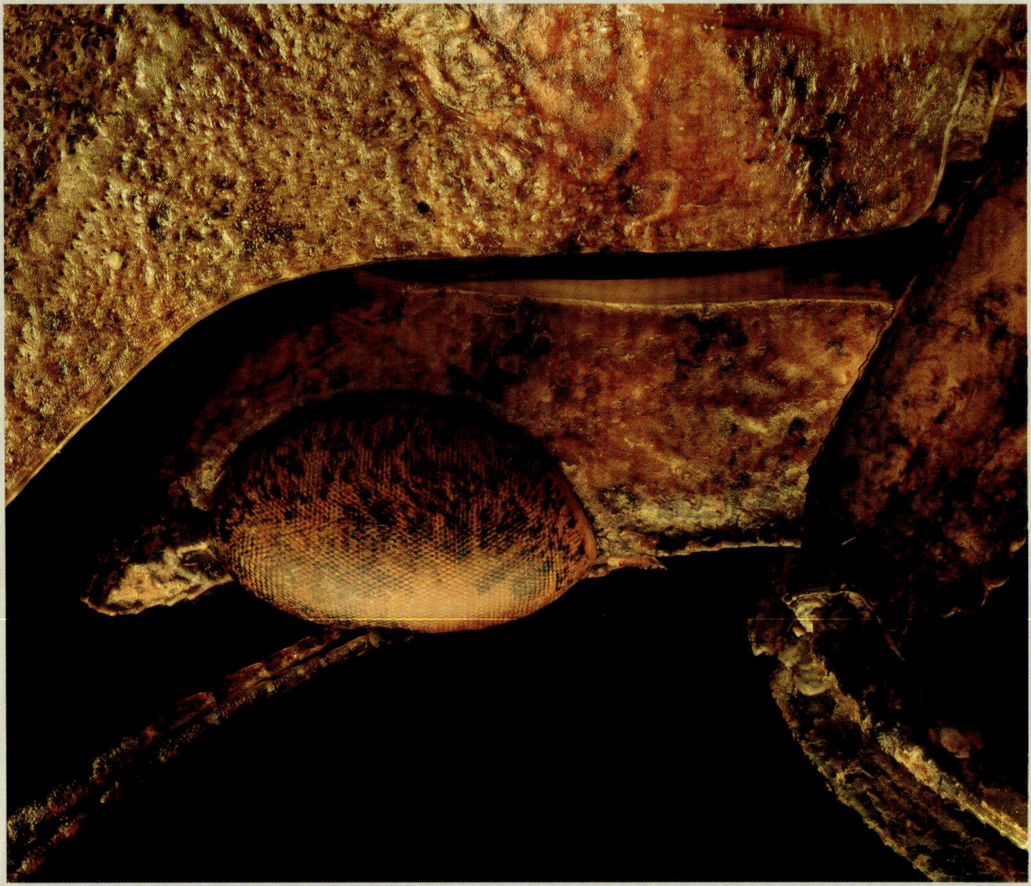
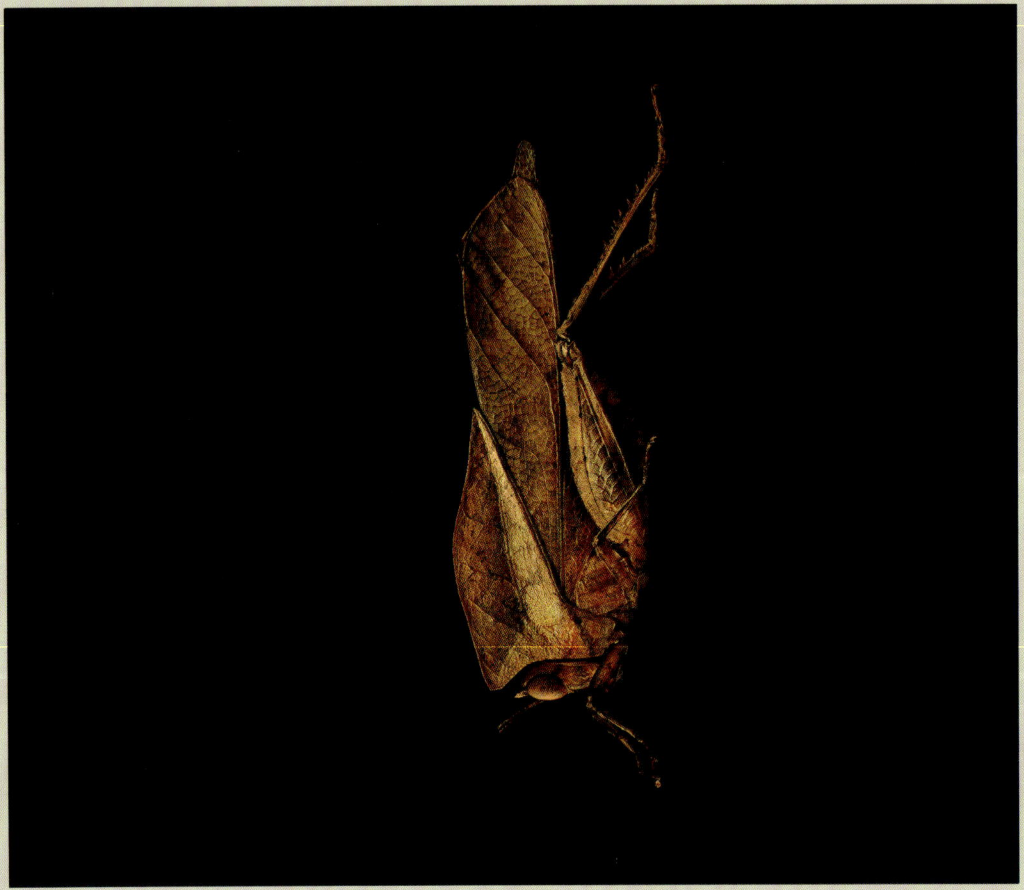
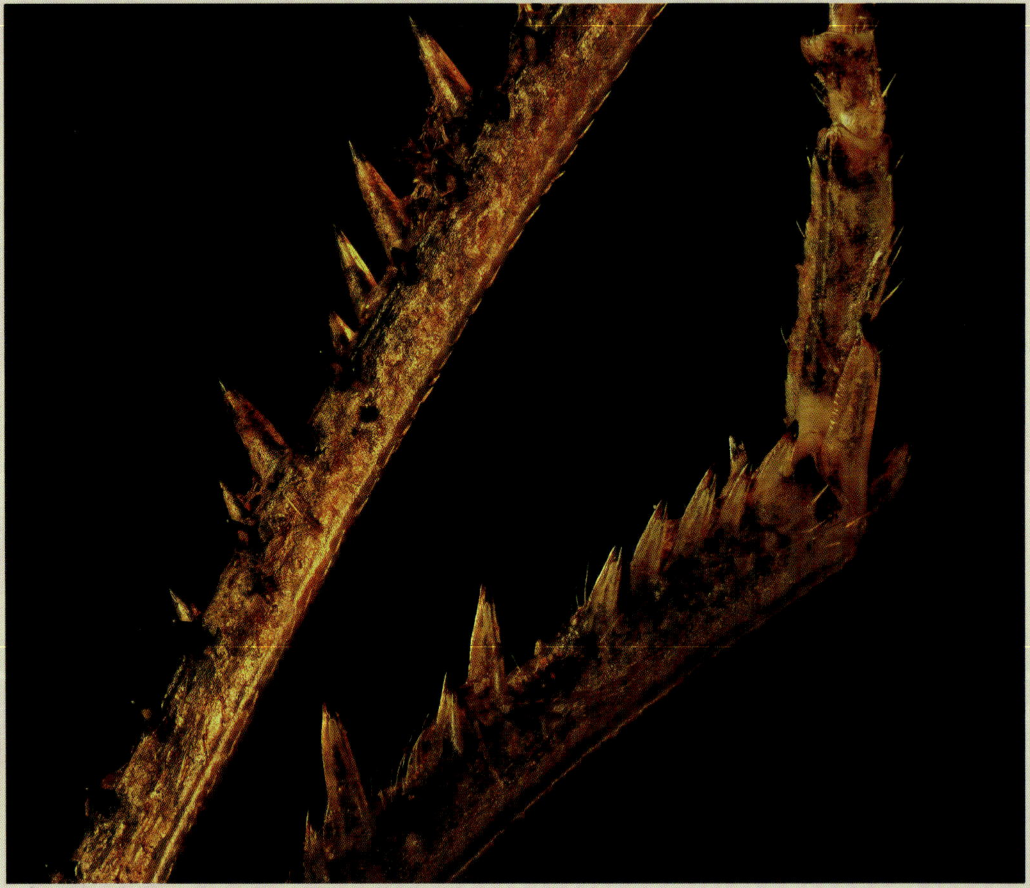

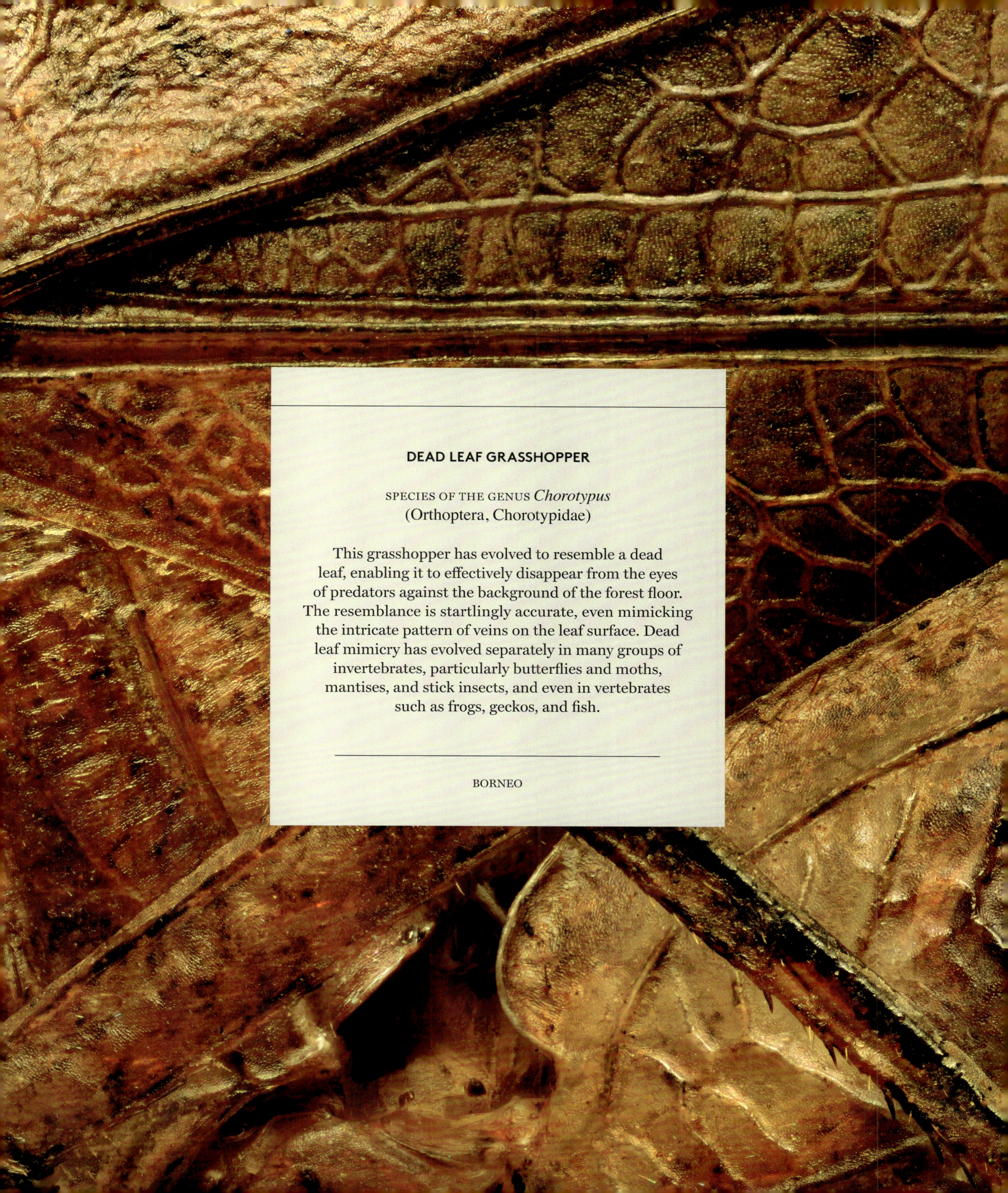

DEAD LEAF GRASSHOPPER

SPECIES OF THE GENUS *Chorotypus*
(Orthoptera, Chorotypidae)

This grasshopper has evolved to resemble a dead leaf, enabling it to effectively disappear from the eyes of predators against the background of the forest floor. The resemblance is startlingly accurate, even mimicking the intricate pattern of veins on the leaf surface. Dead leaf mimicry has evolved separately in many groups of invertebrates, particularly butterflies and moths, mantises, and stick insects, and even in vertebrates such as frogs, geckos, and fish.

BORNEO

PARIS PEACOCK

Papilio paris (Lepidoptera, Papilionidae)

This butterfly belongs to a group known as peacock swallowtails, inhabitants of the tropical forests of Southeast Asia and Australasia. They exist as a great many subspecies and island forms, some of which are now very rare due to habitat loss. Peacock swallowtails often share a characteristic wing pattern of delicate scatterings of intense green scales against a background of brown or black.

INDONESIA

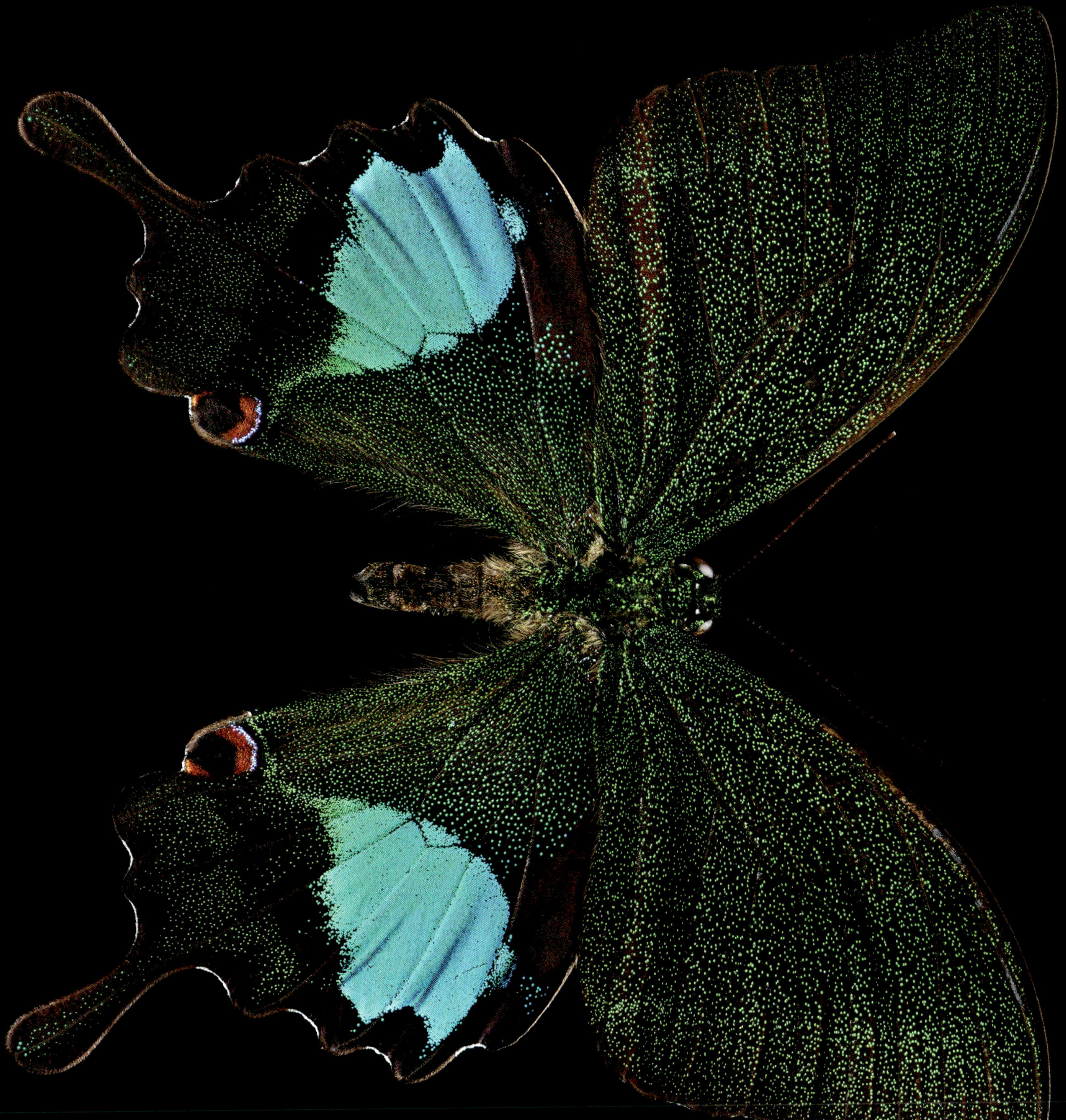

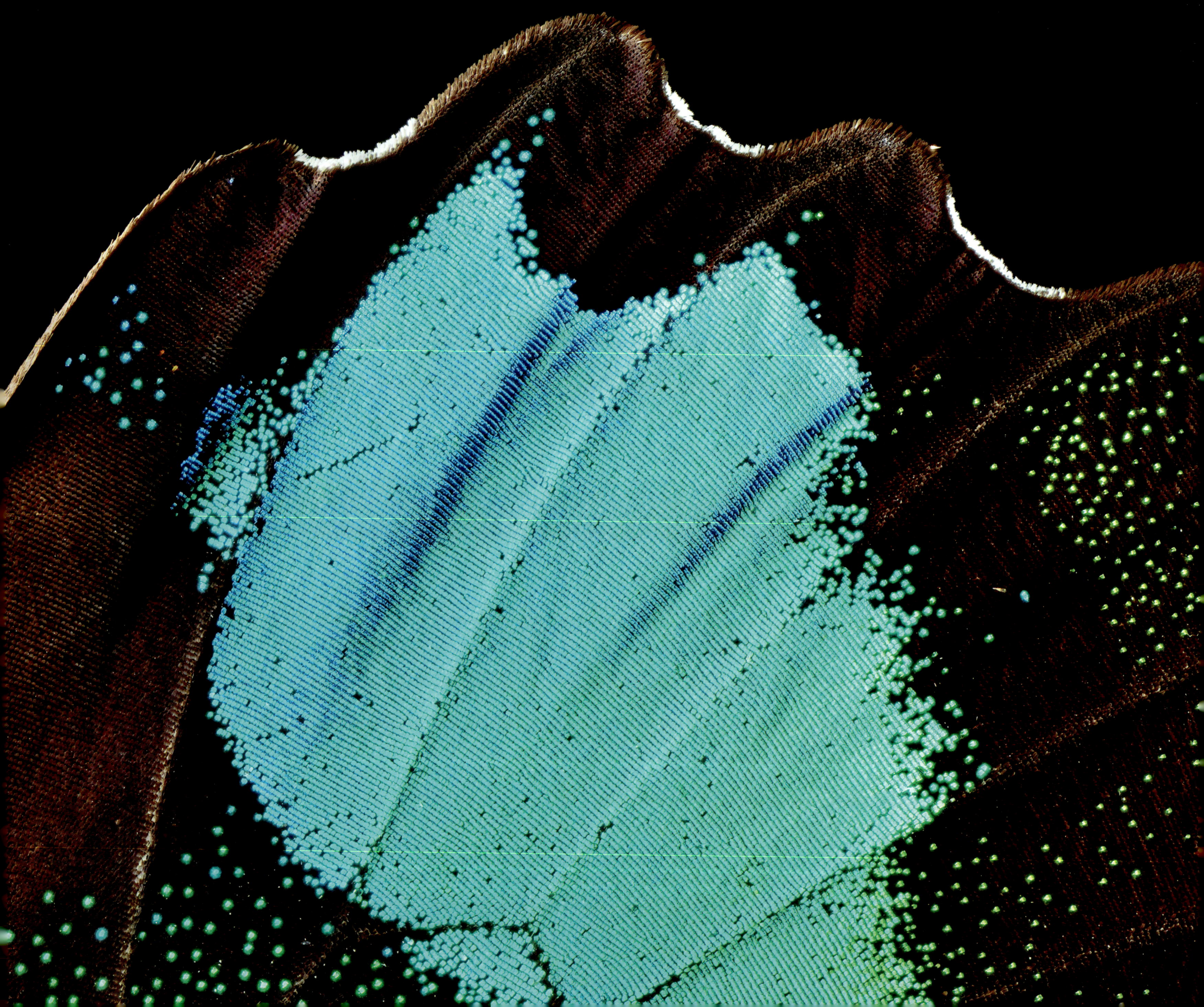

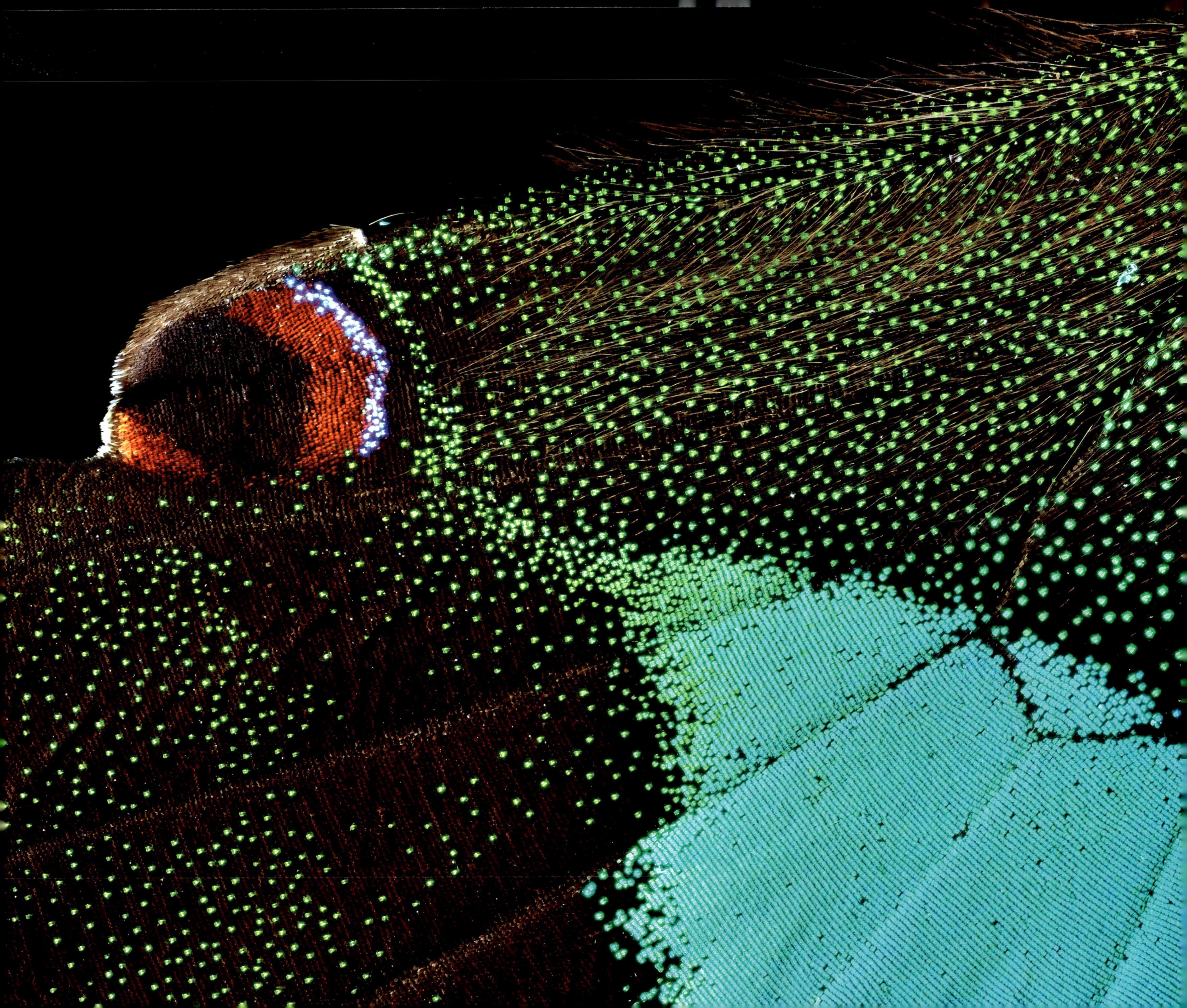

POTTER WASP

SPECIES OF THE SUBFAMILY *Eumeninae*
(Hymenoptera, Vespidae)

At high magnification some interesting anatomical details of this potter wasp become apparent. The tips of the antennae of this male wasp are modified into hooks, used to hold the antennae of the female during mating. Covering large sections of the wings, extremely small hairs, or macrotrichia, likely have an aerodynamic function, influencing airflow over the wing in flight.

INDIA

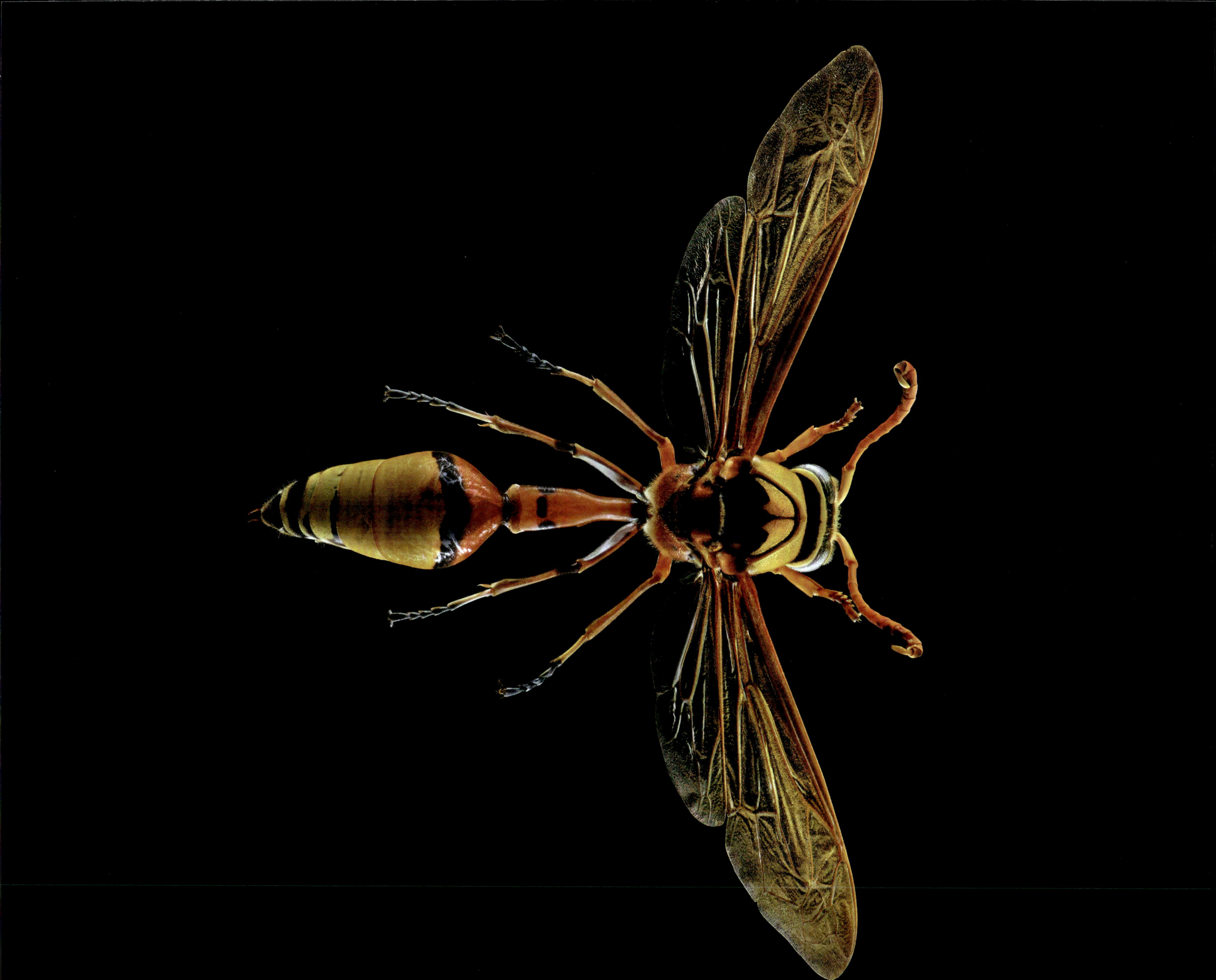

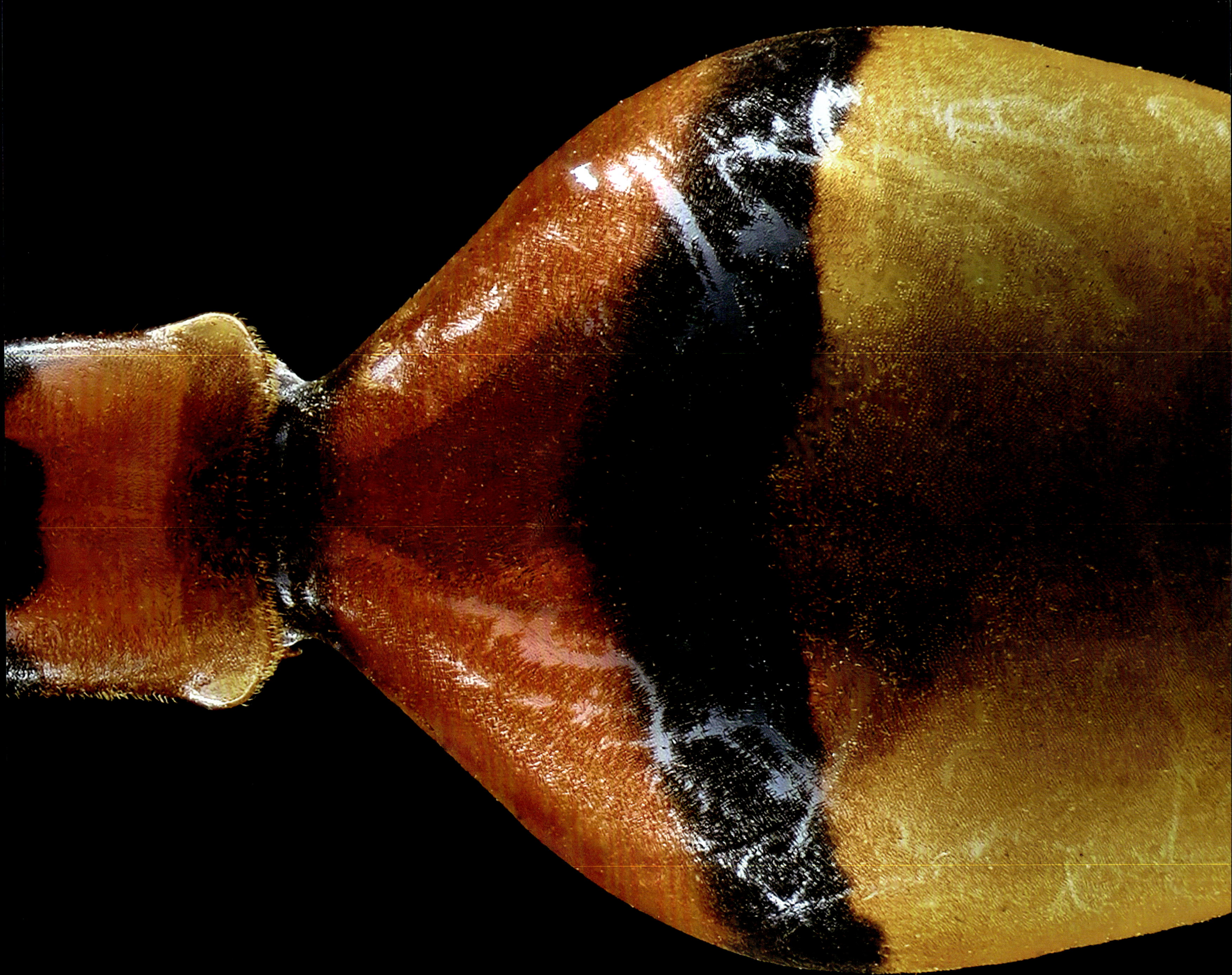

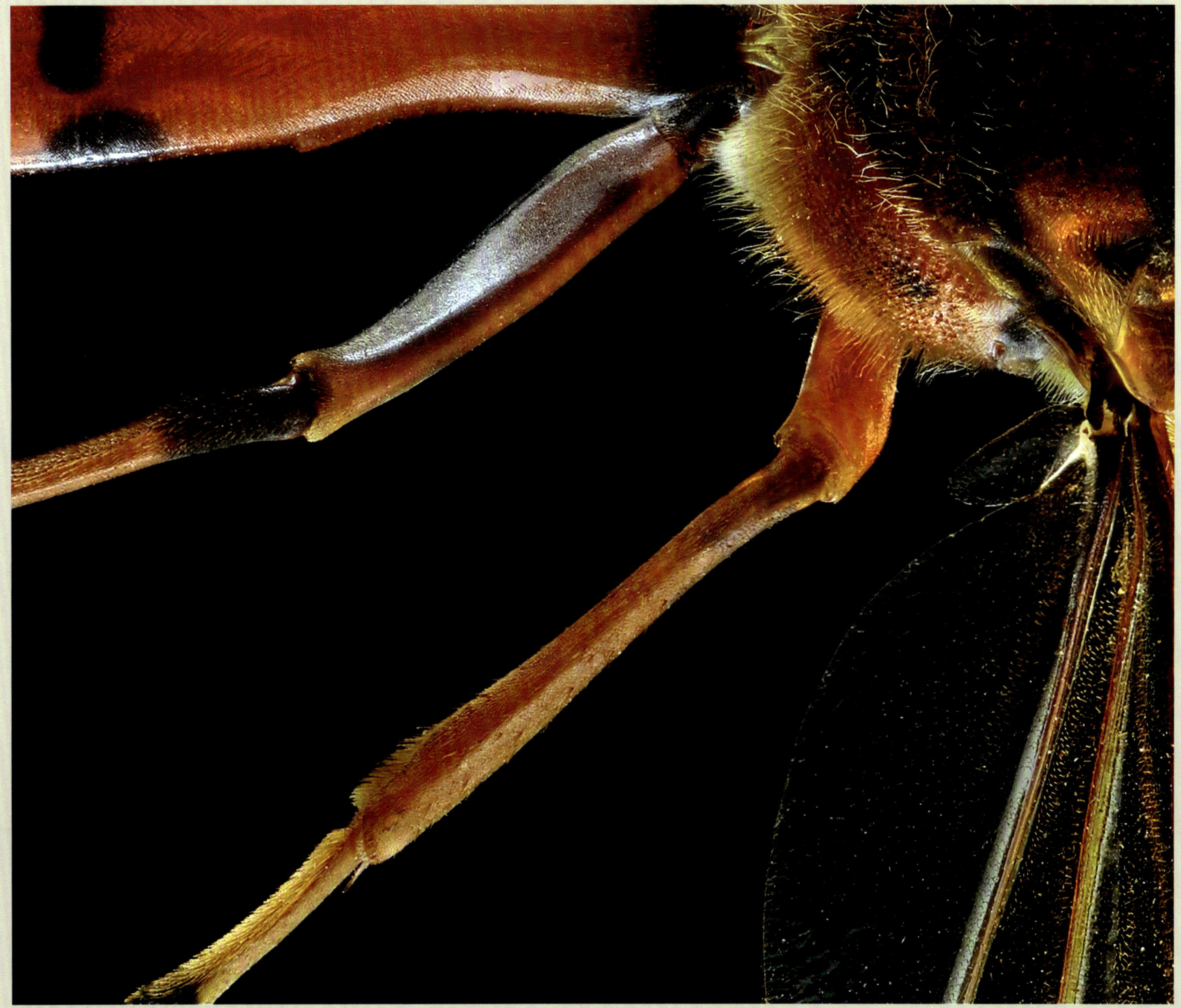

RUBY-TAILED WASP

Parnopes grandior (Hymenoptera, Chrysididae)

Ruby-tailed wasps are famed for their bright colors. Also known as jewel or cuckoo wasps, most species are parasites of other solitary wasps or bees. Unlike most of their relatives, ruby-tailed wasps are unable to sting. To protect themselves against attacks, they have developed an extremely tough armored exoskeleton and many species are able to curl up into a ball to protect their more delicate legs and head. This specimen shows a very interesting surface sculpture of different-sized pits covering the entire body.

FRANCE

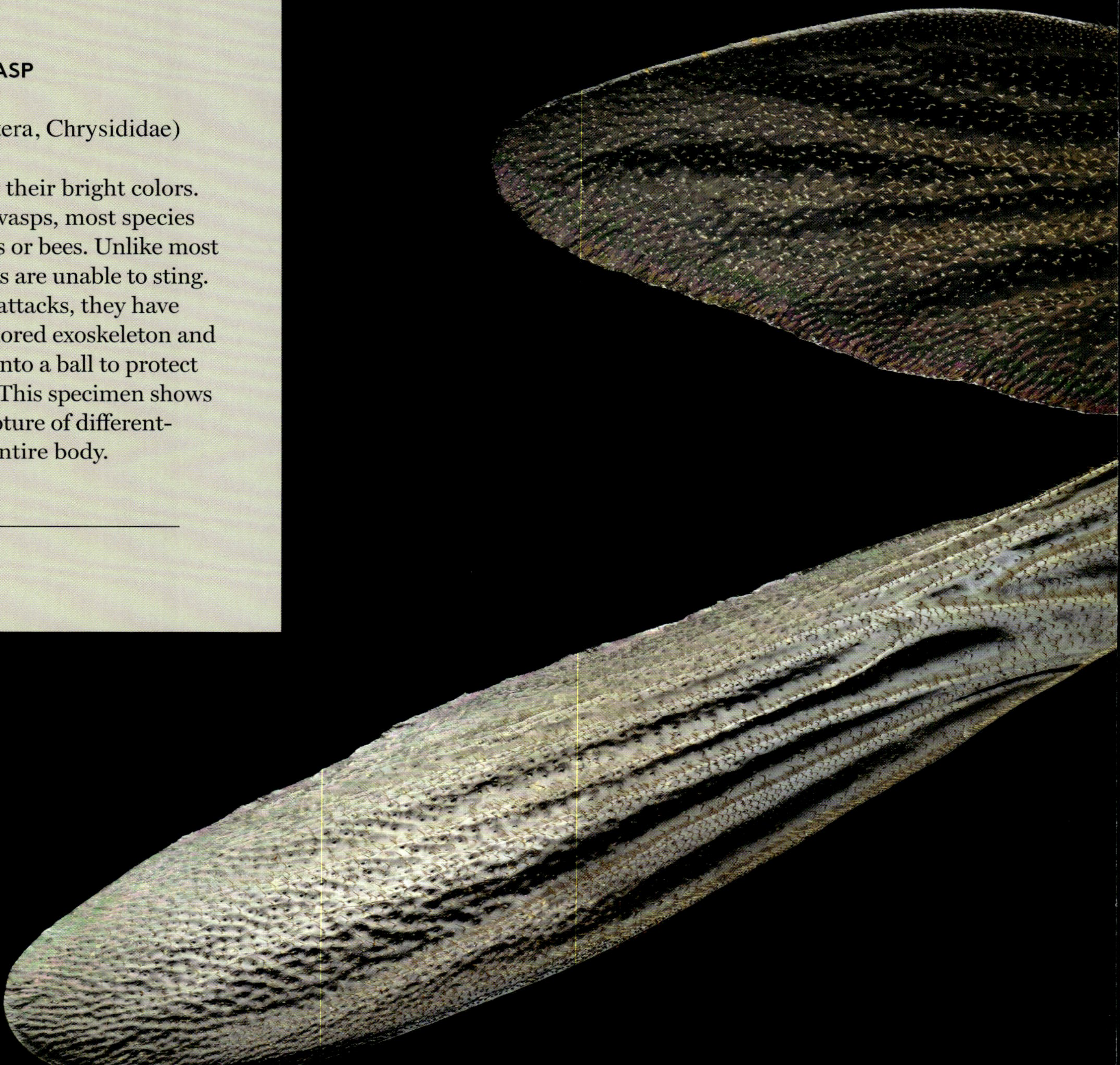

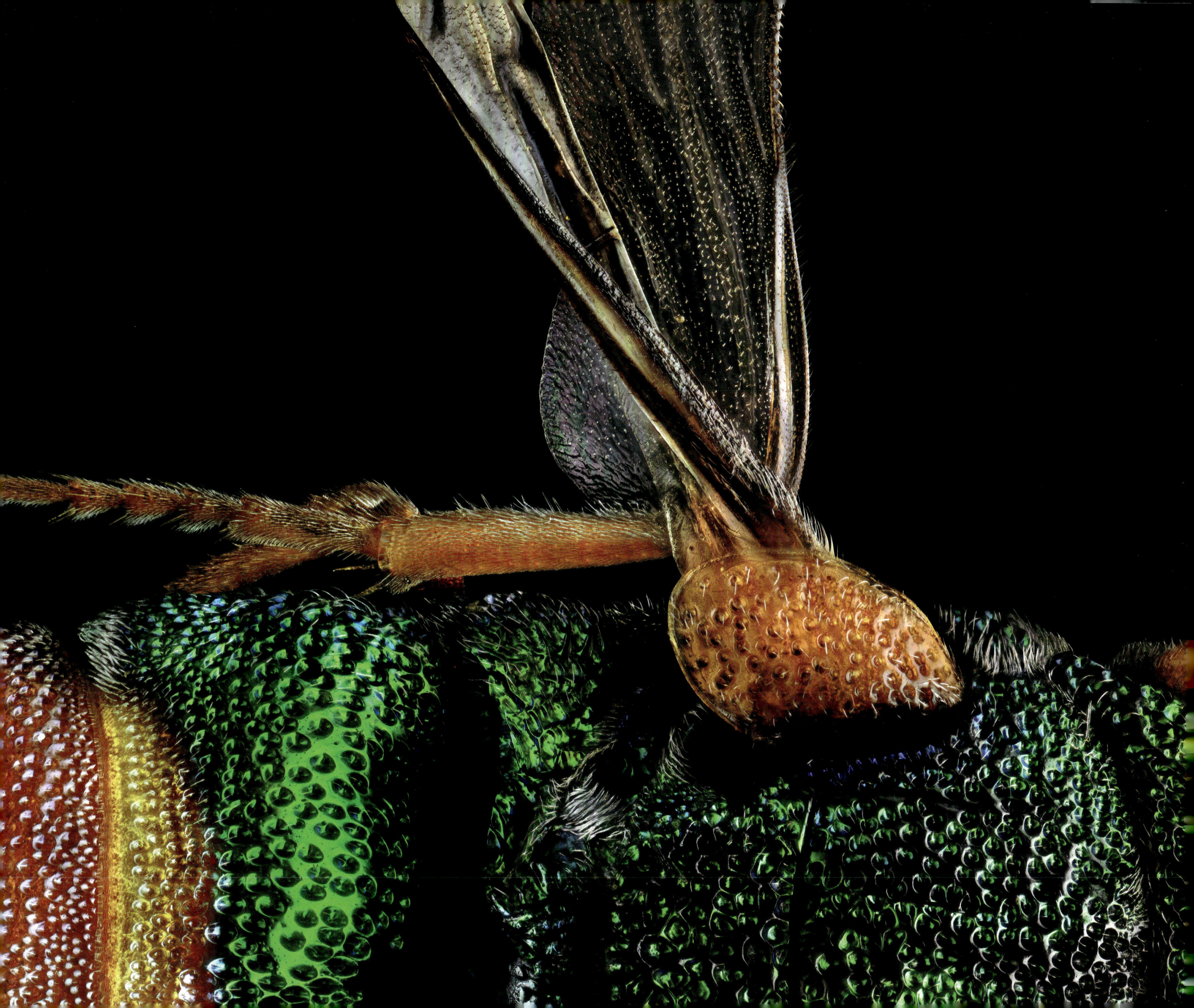

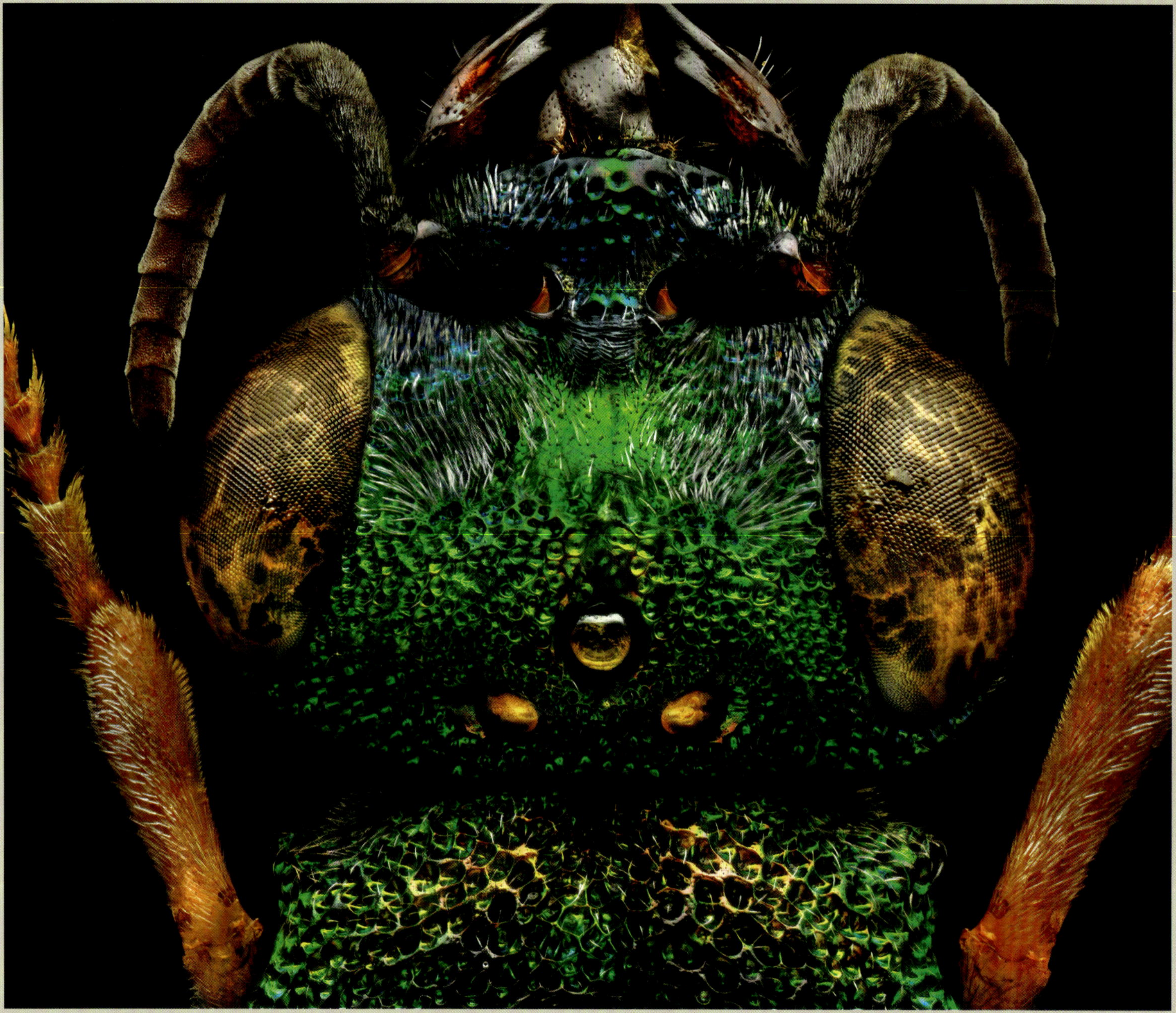

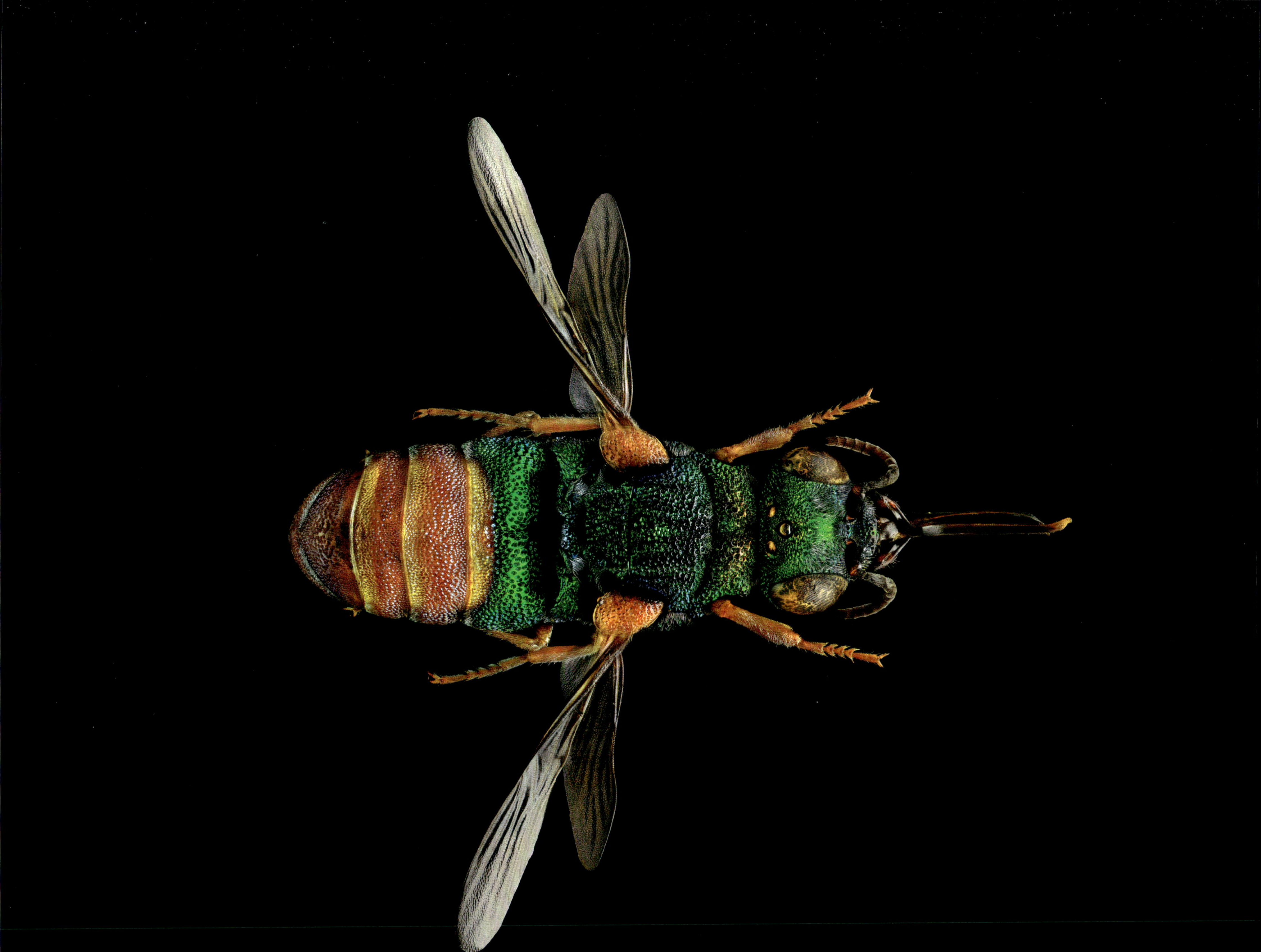

PLEASING FUNGUS BEETLE

SPECIES OF THE GENUS *Brachysphaenus*
(Coleoptera, Erotylidae)

Close relatives of the ladybugs, pleasing fungus beetles often have conspicuous markings in various combinations of bright colors, spots, stripes, and other patterns. Their attractive colors advertise a sophisticated chemical defense system to predators. Noxious chemicals are delivered both by glands on the surface of the body and also by reflex bleeding, where the insect exudes a mixture of chemicals and its own blood from the end of its abdomen.

BOLIVIA

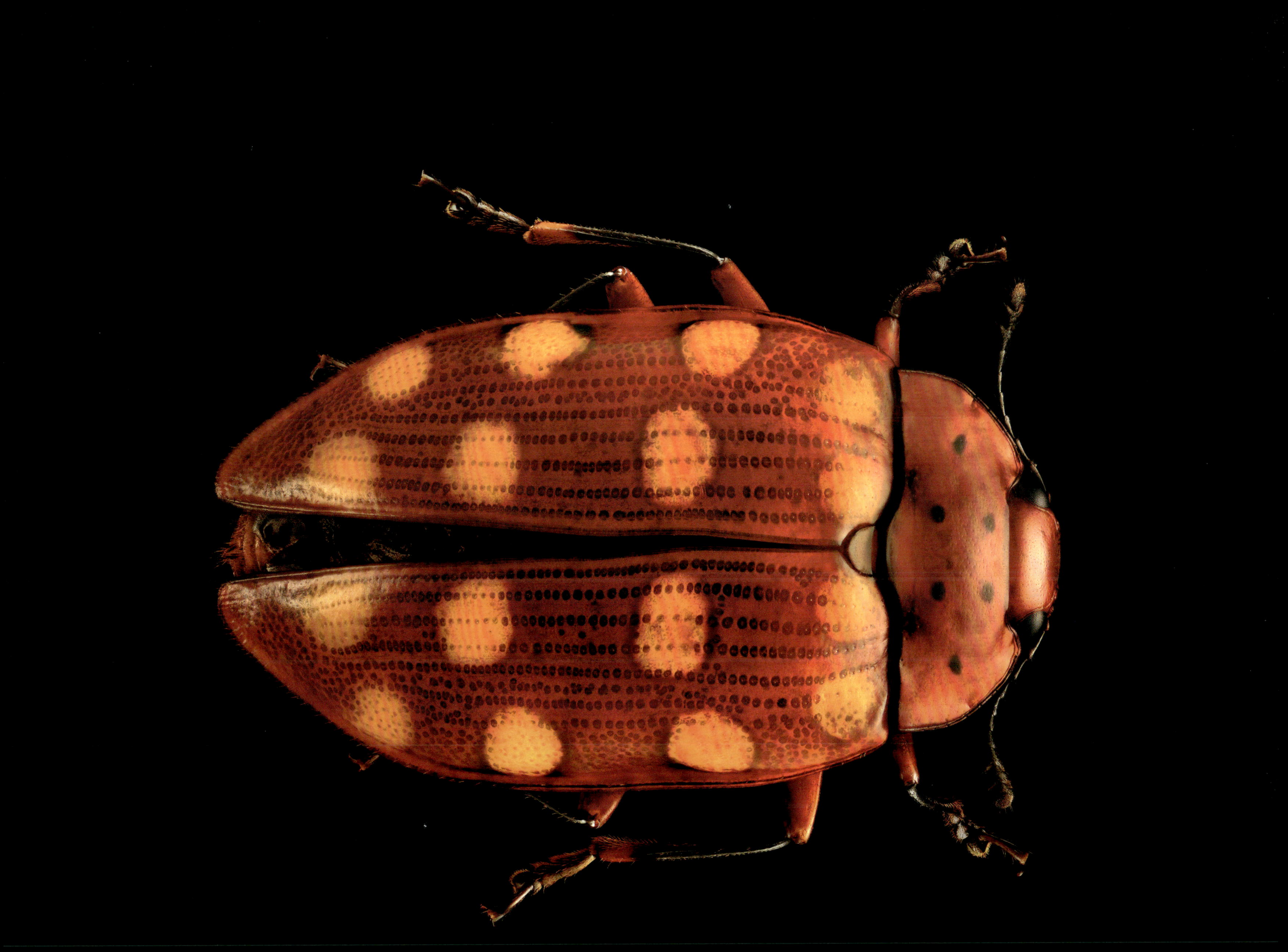

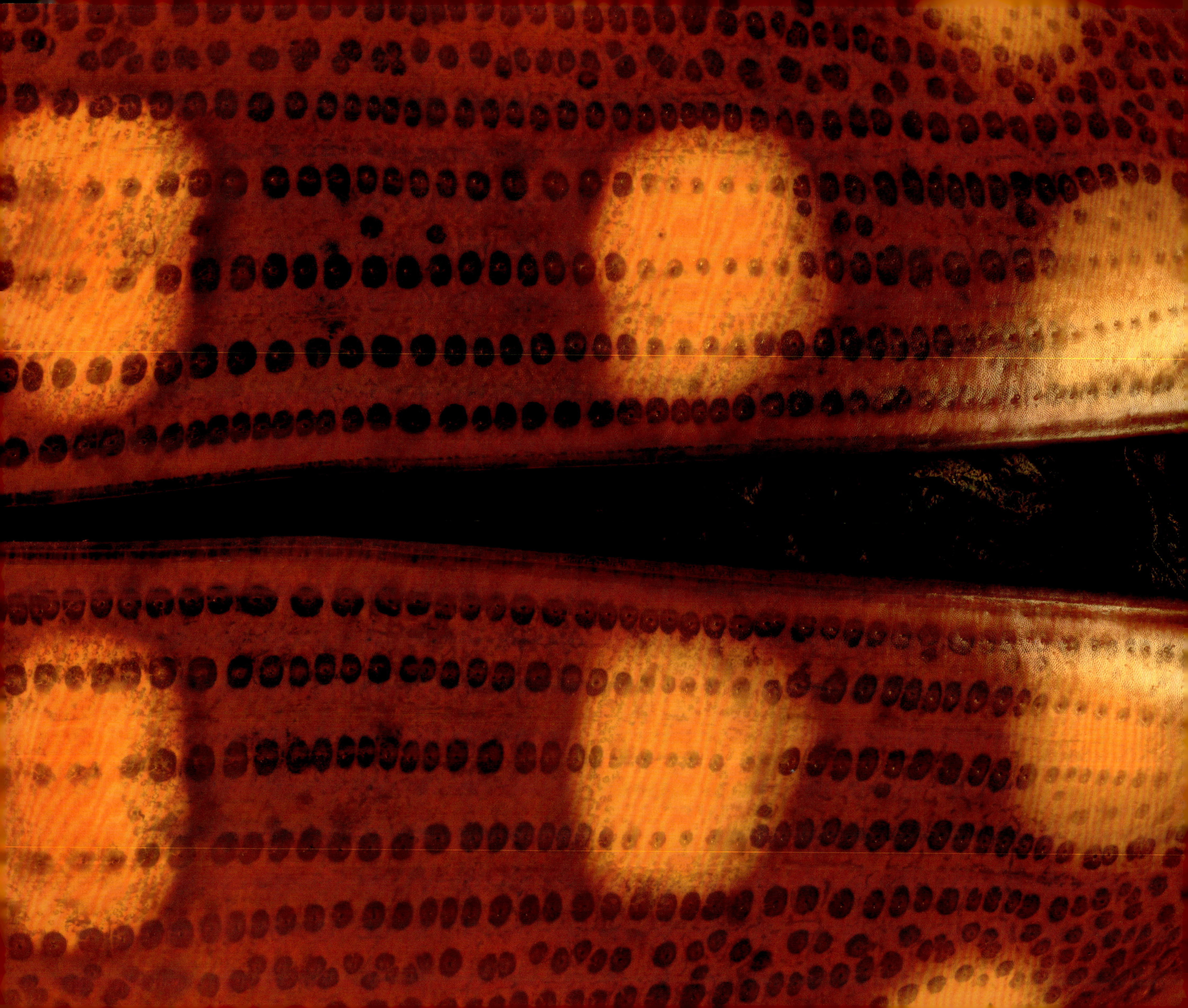

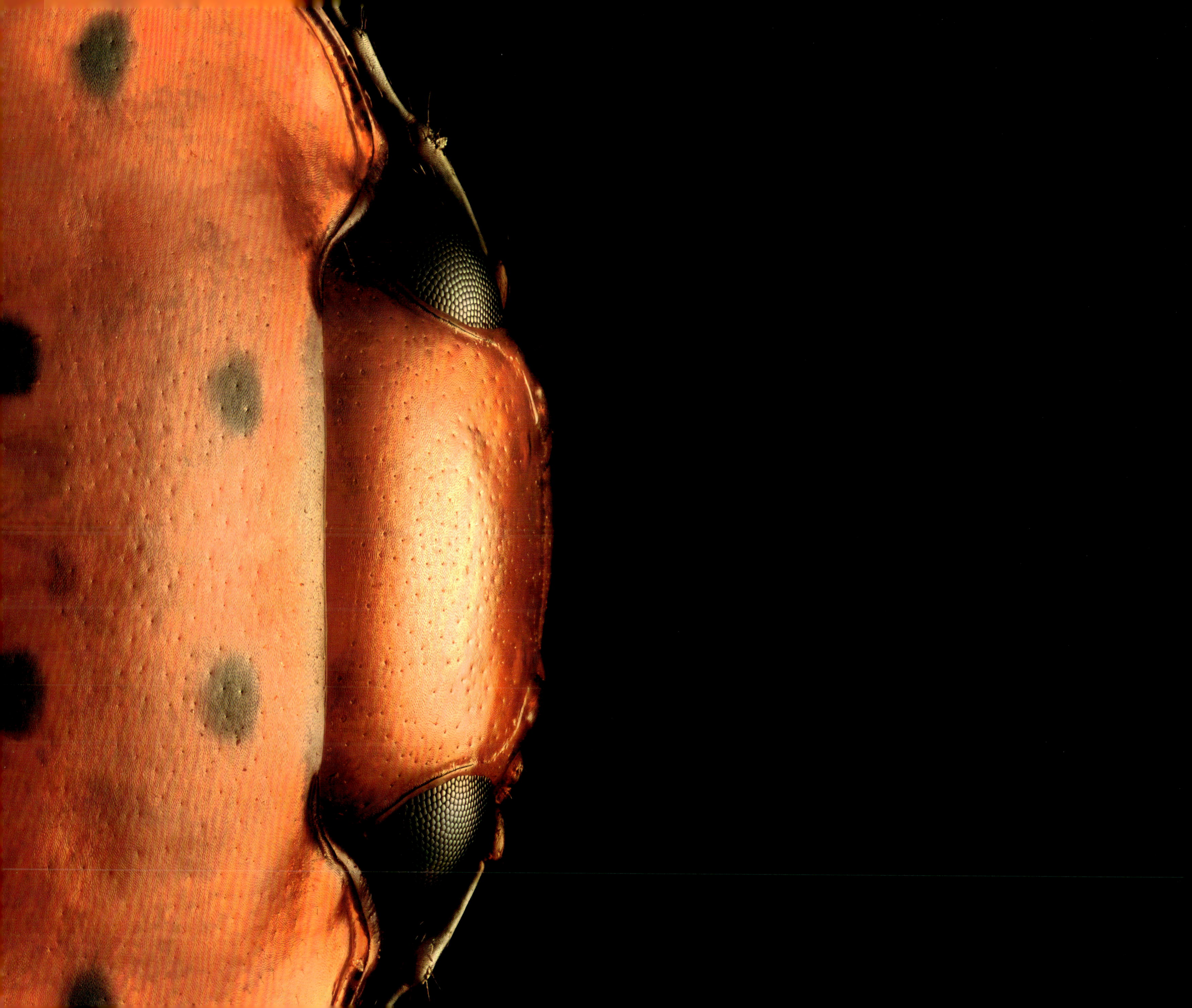

MARION FLIGHTLESS MOTH

Pringleophaga marioni (Lepidoptera, Tineidae)

At first sight it is hard to tell exactly which group this insect belongs to, but at high magnification the flattened scales covering the legs and body reveal its true identity—a species of moth. This moth is unique to the Prince Edward Islands, two cold, remote, and windswept islands in the sub-Antarctic Indian Ocean, more than a thousand miles from South Africa. It has a highly specialized lifestyle, feeding on the detritus accumulated in albatross nests. Although completely flightless, its wings have not been lost entirely and are reduced to short, narrow structures.

MARION ISLAND

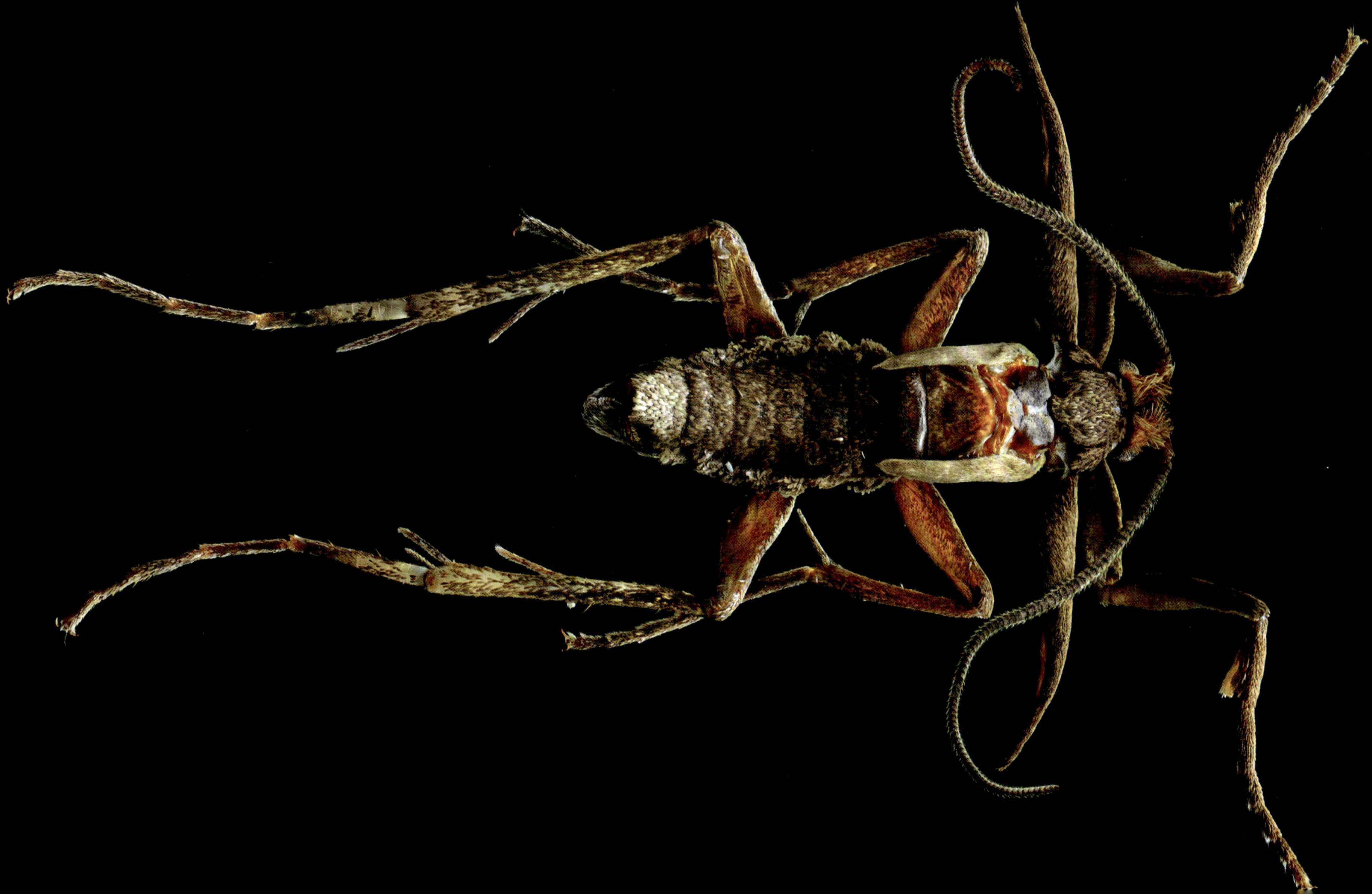

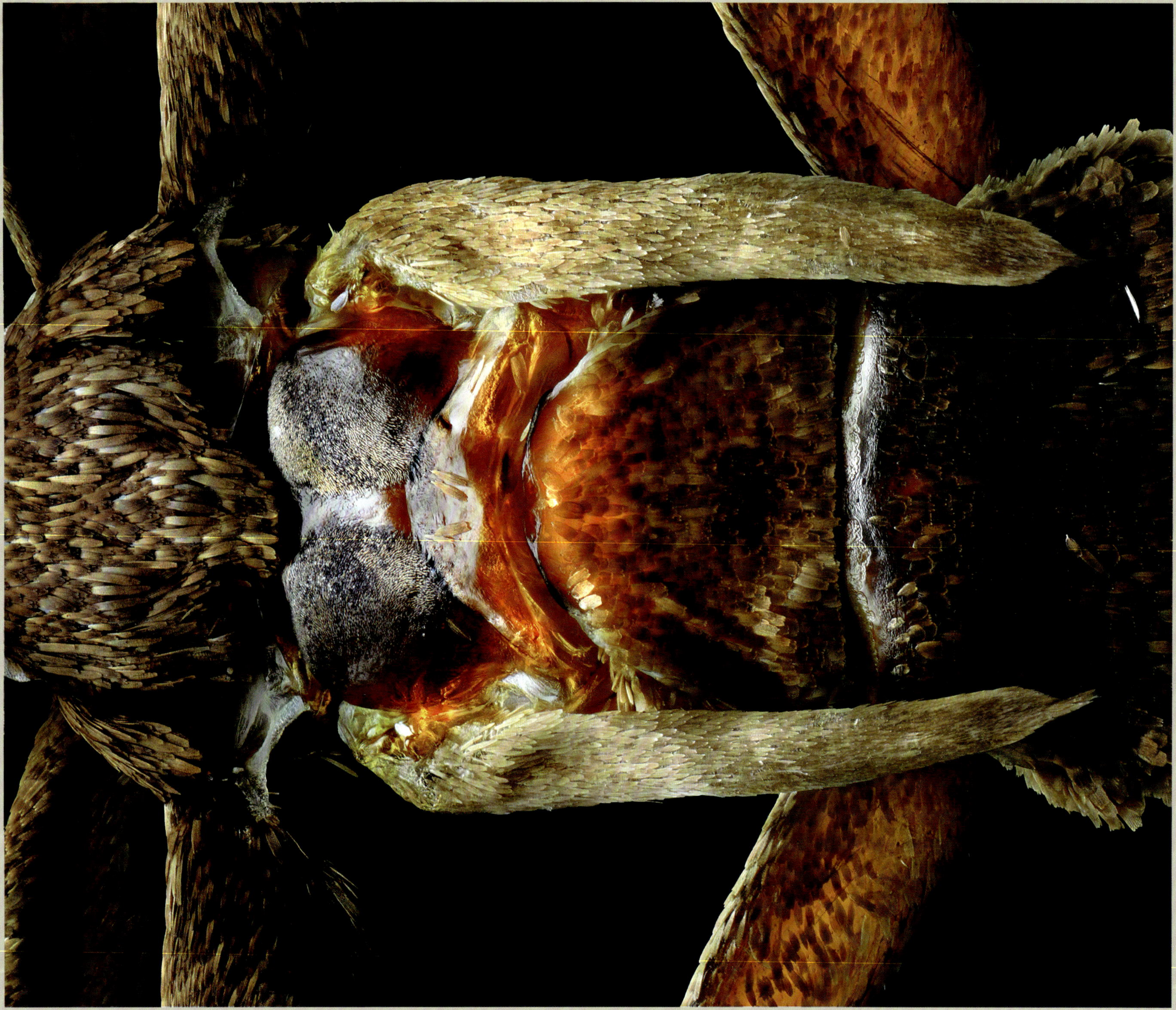

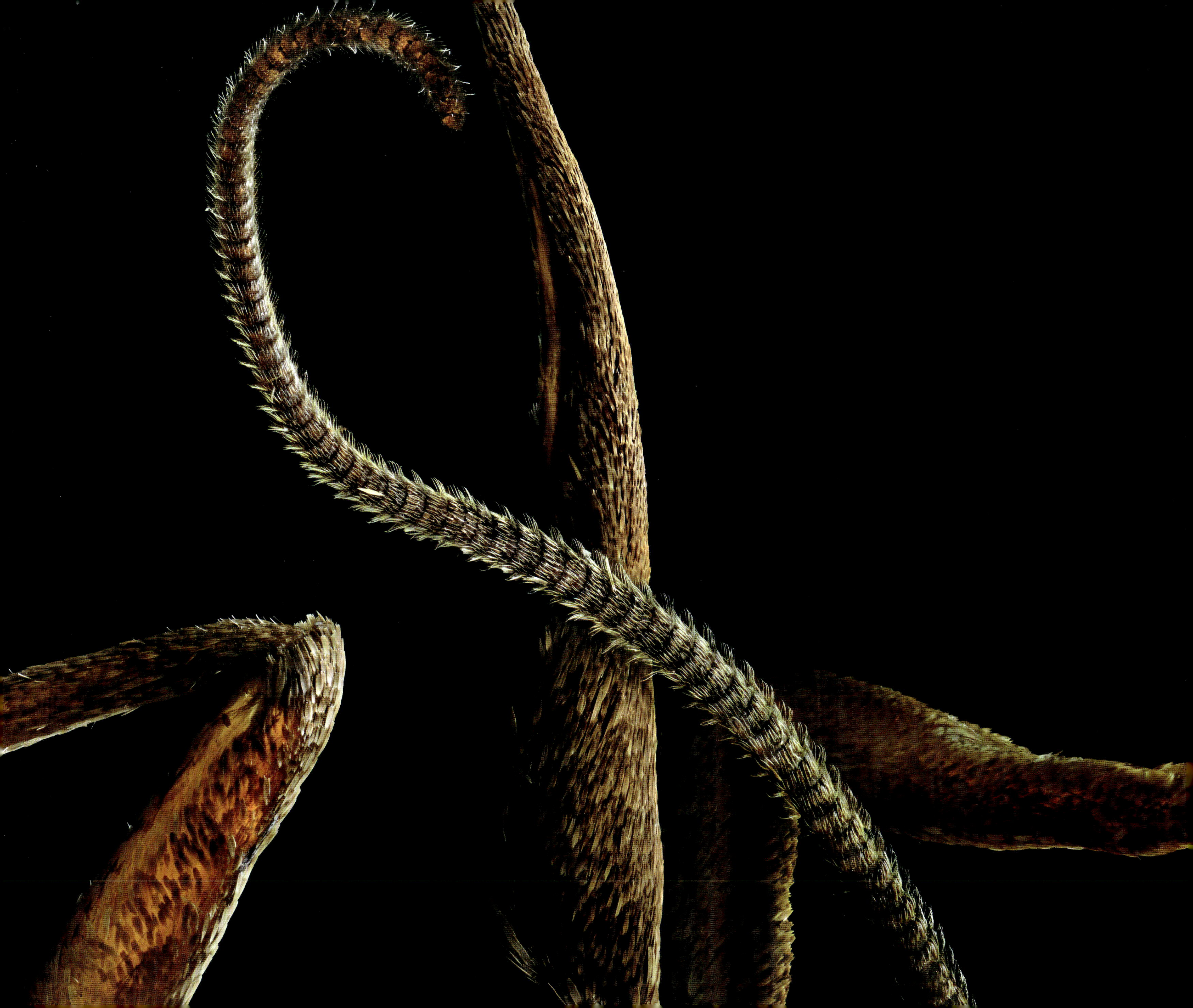

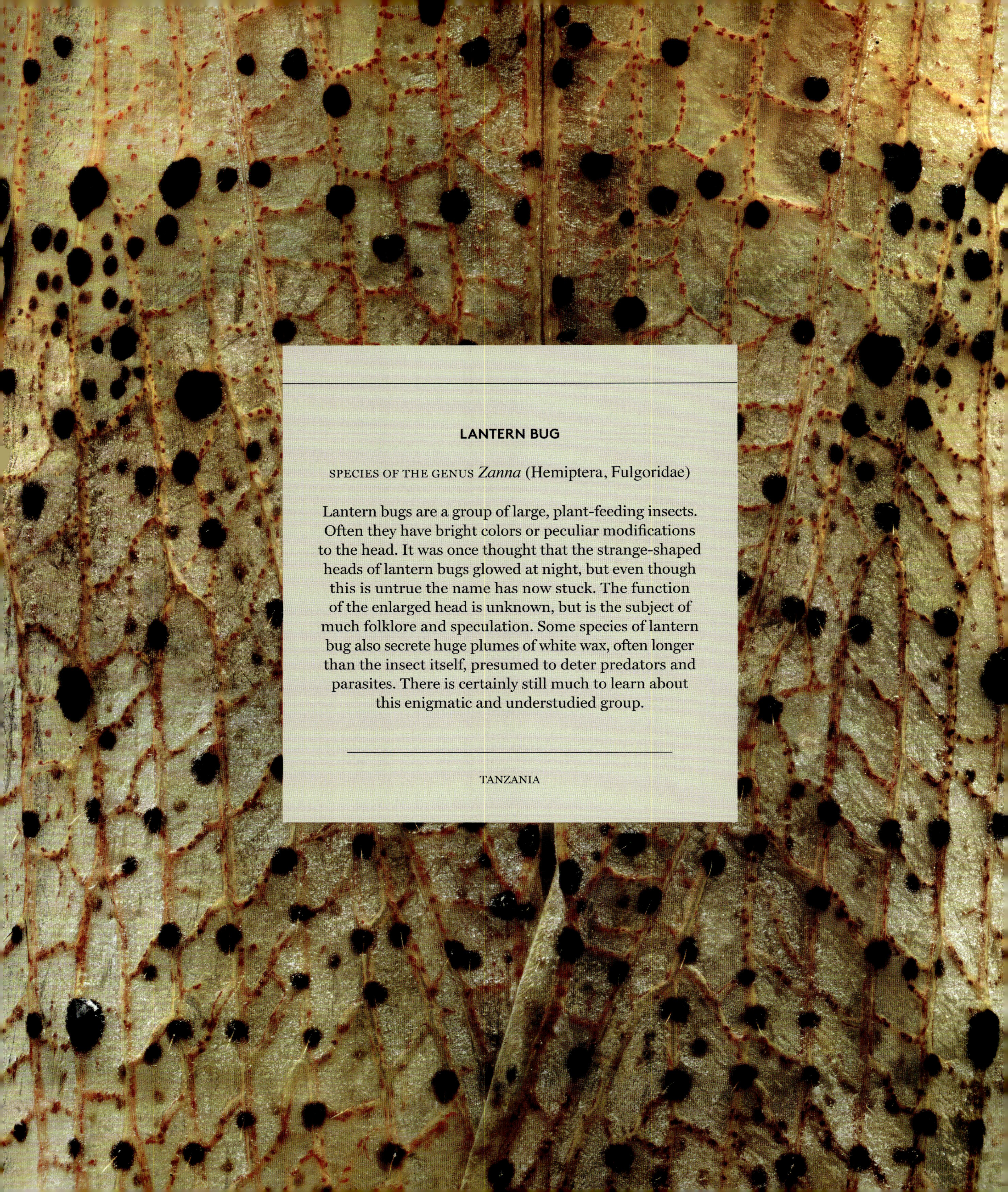

LANTERN BUG

SPECIES OF THE GENUS *Zanna* (Hemiptera, Fulgoridae)

Lantern bugs are a group of large, plant-feeding insects. Often they have bright colors or peculiar modifications to the head. It was once thought that the strange-shaped heads of lantern bugs glowed at night, but even though this is untrue the name has now stuck. The function of the enlarged head is unknown, but is the subject of much folklore and speculation. Some species of lantern bug also secrete huge plumes of white wax, often longer than the insect itself, presumed to deter predators and parasites. There is certainly still much to learn about this enigmatic and understudied group.

TANZANIA

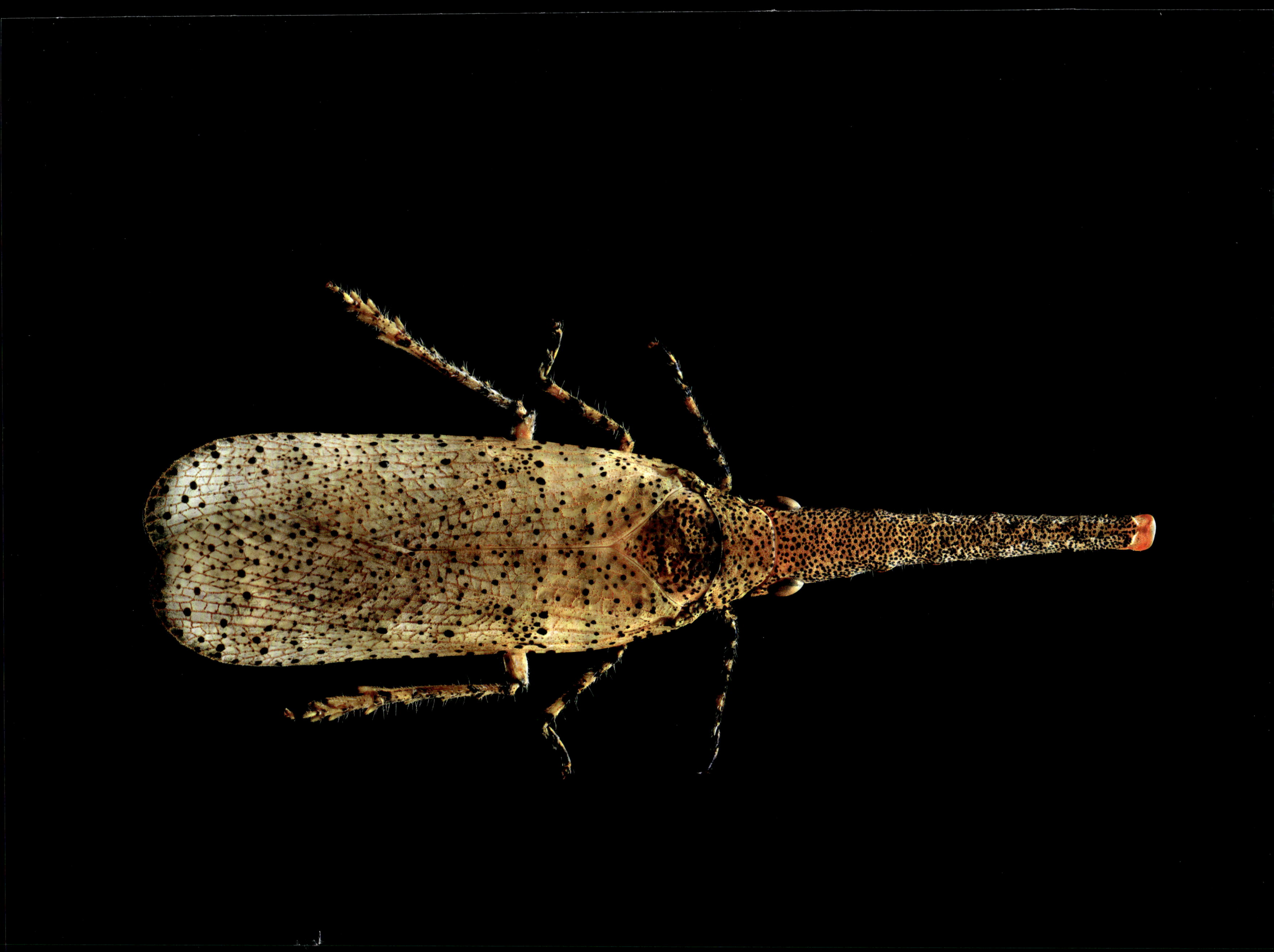

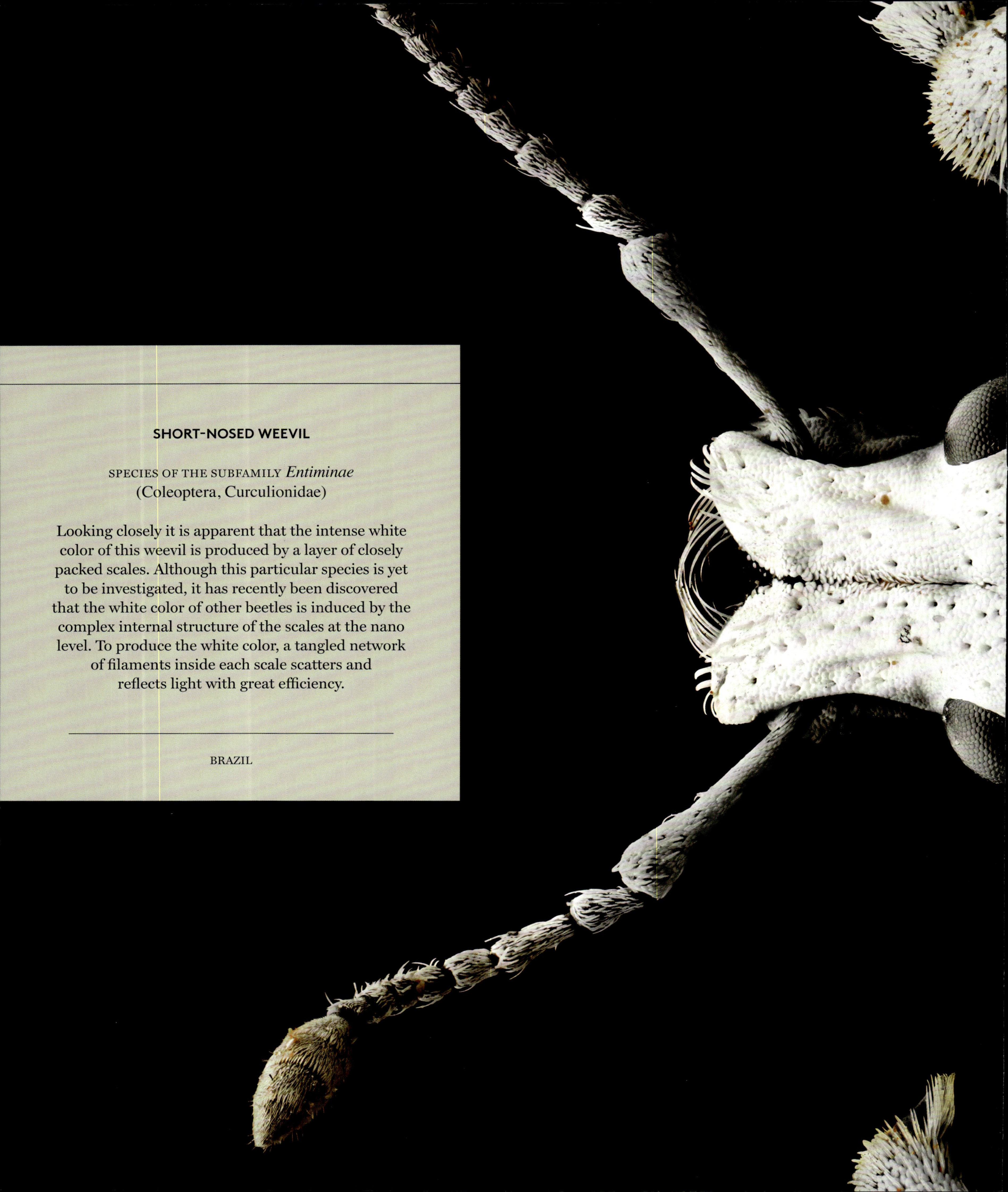

SHORT-NOSED WEEVIL

SPECIES OF THE SUBFAMILY *Entiminae*
(Coleoptera, Curculionidae)

Looking closely it is apparent that the intense white color of this weevil is produced by a layer of closely packed scales. Although this particular species is yet to be investigated, it has recently been discovered that the white color of other beetles is induced by the complex internal structure of the scales at the nano level. To produce the white color, a tangled network of filaments inside each scale scatters and reflects light with great efficiency.

BRAZIL

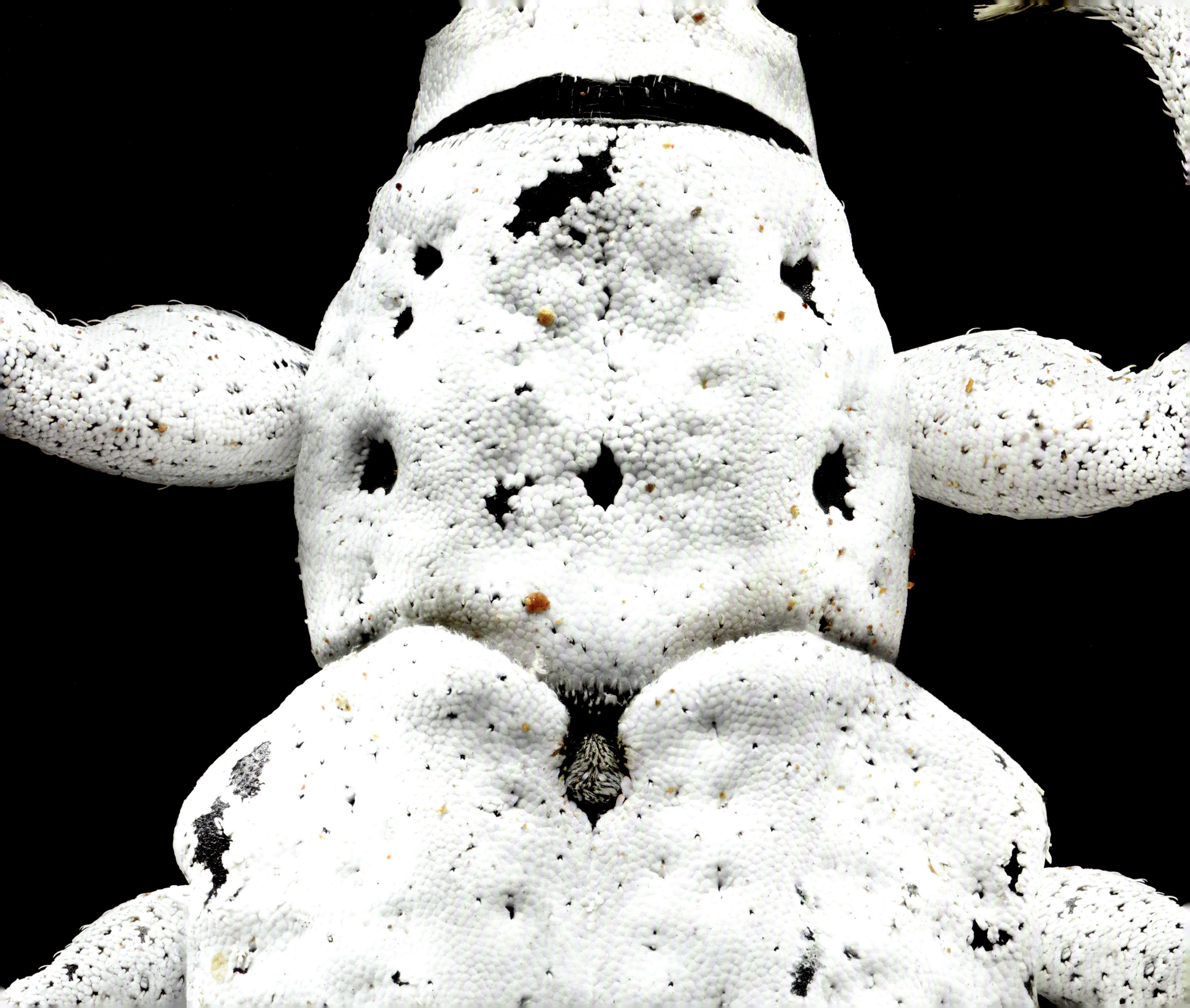

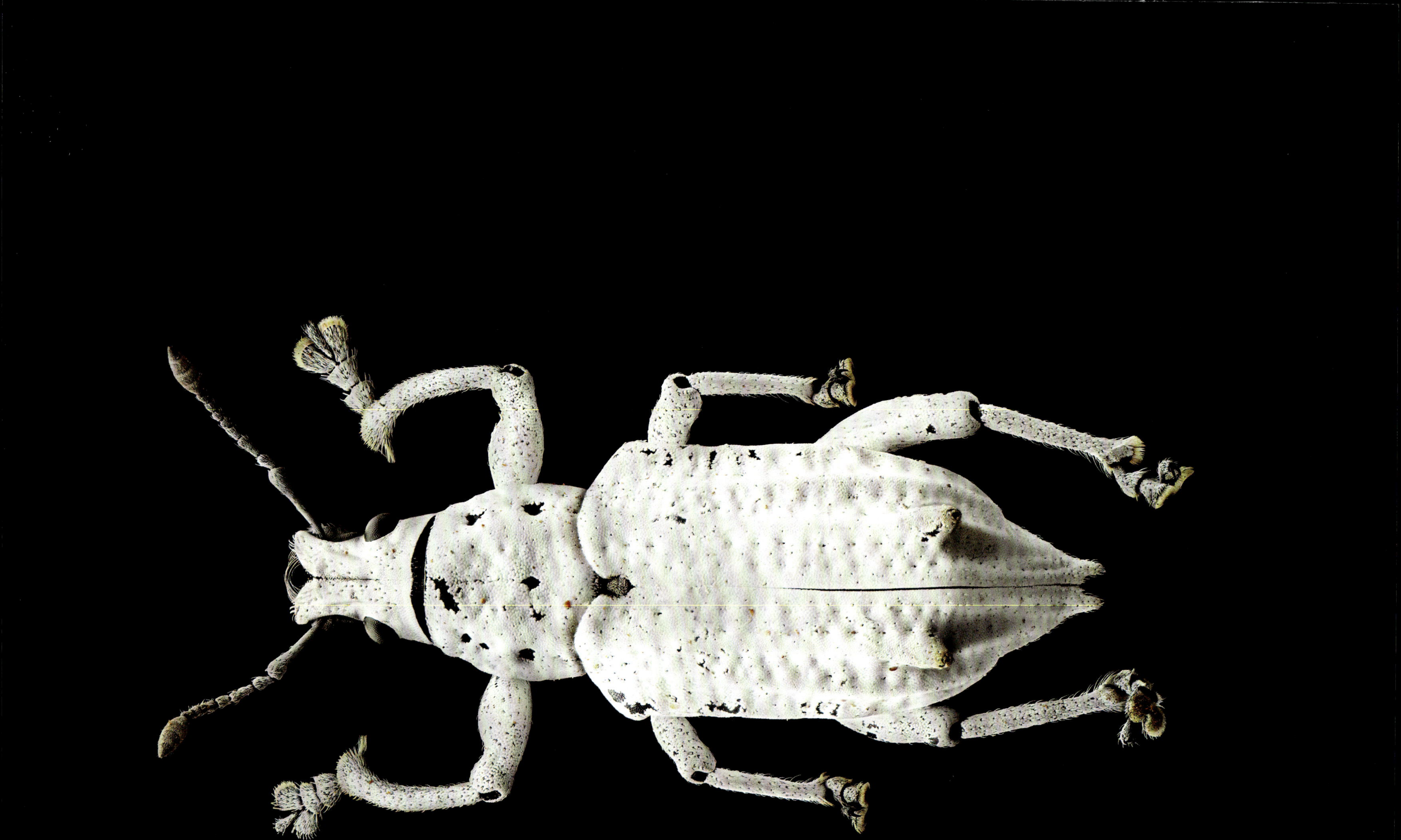

AMAZONIAN PURPLE WARRIOR SCARAB

Coprophanaeus lancifer (Coleoptera, Scarabaeidae)

A large and impressive scarab beetle found widely across the Amazon Basin, this species is particularly active at dusk when the contrast in color between the black horn and the blue body is enhanced, helping individuals to recognize others of the same species. Although belonging to a group of mainly dung-feeding scarabs, this and related species have switched to feeding on dead animals. The toothed, blade-like area at the front of the head and the serrated front legs are thought to help in breaking up tough carrion.

PERU

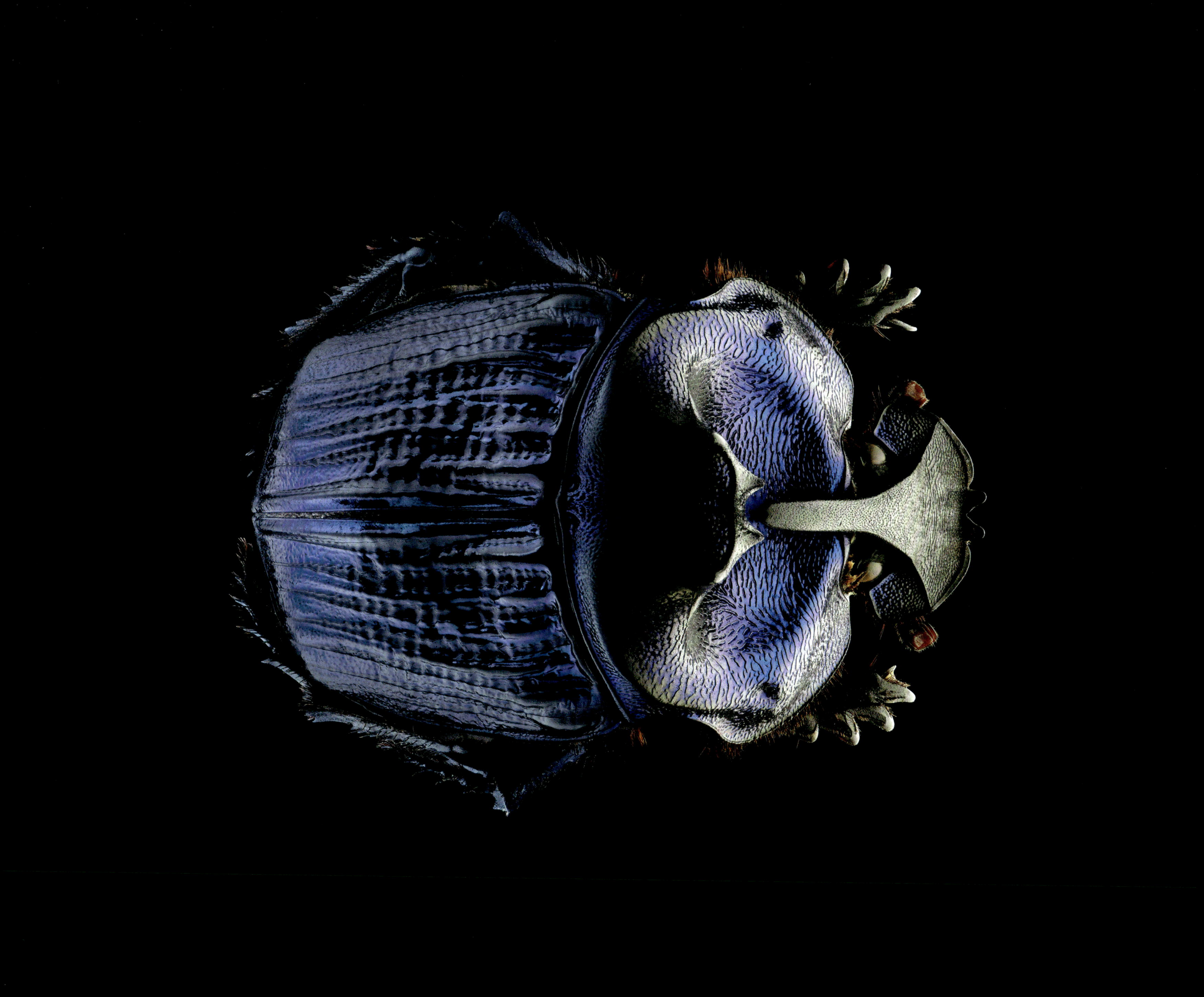

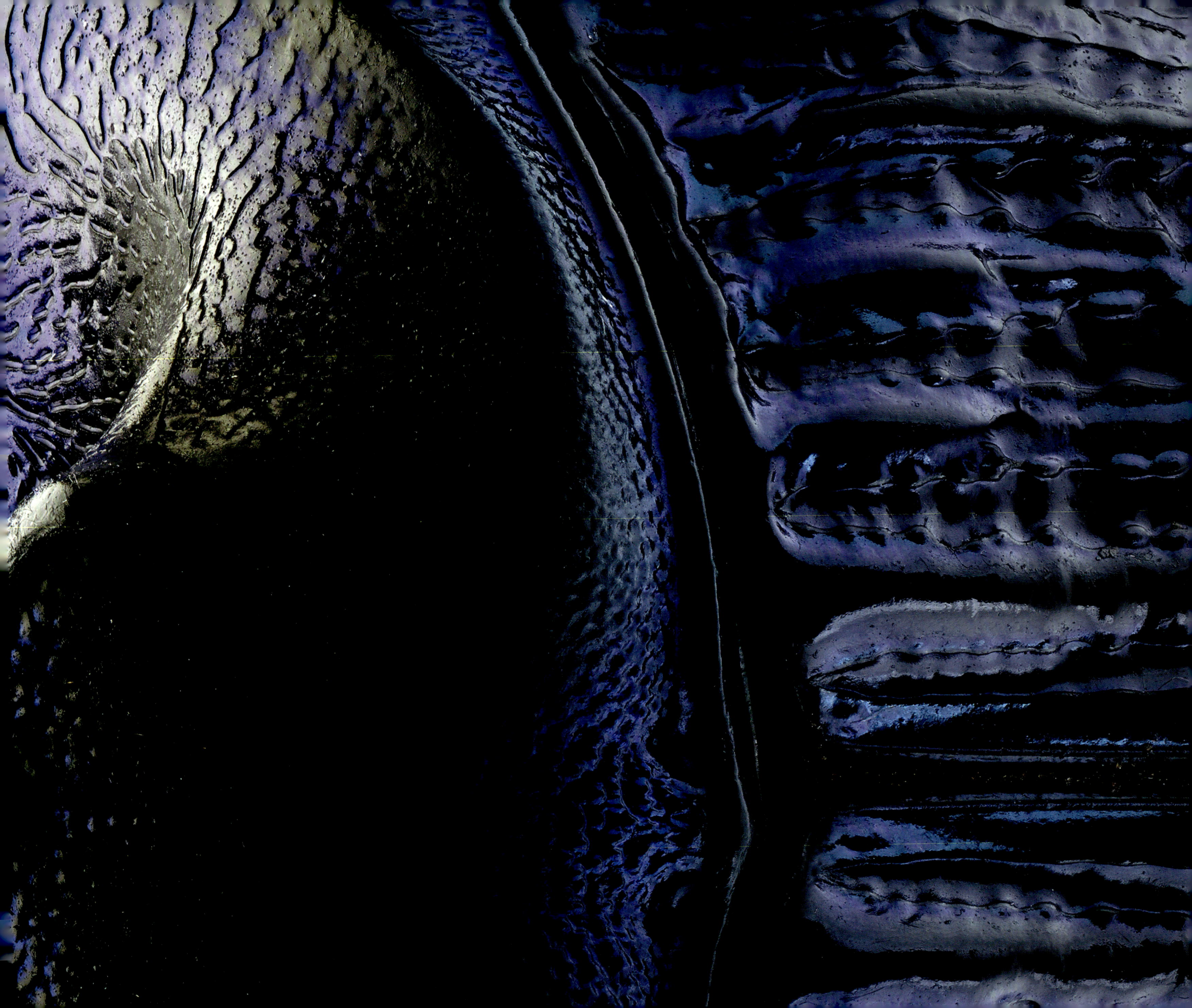

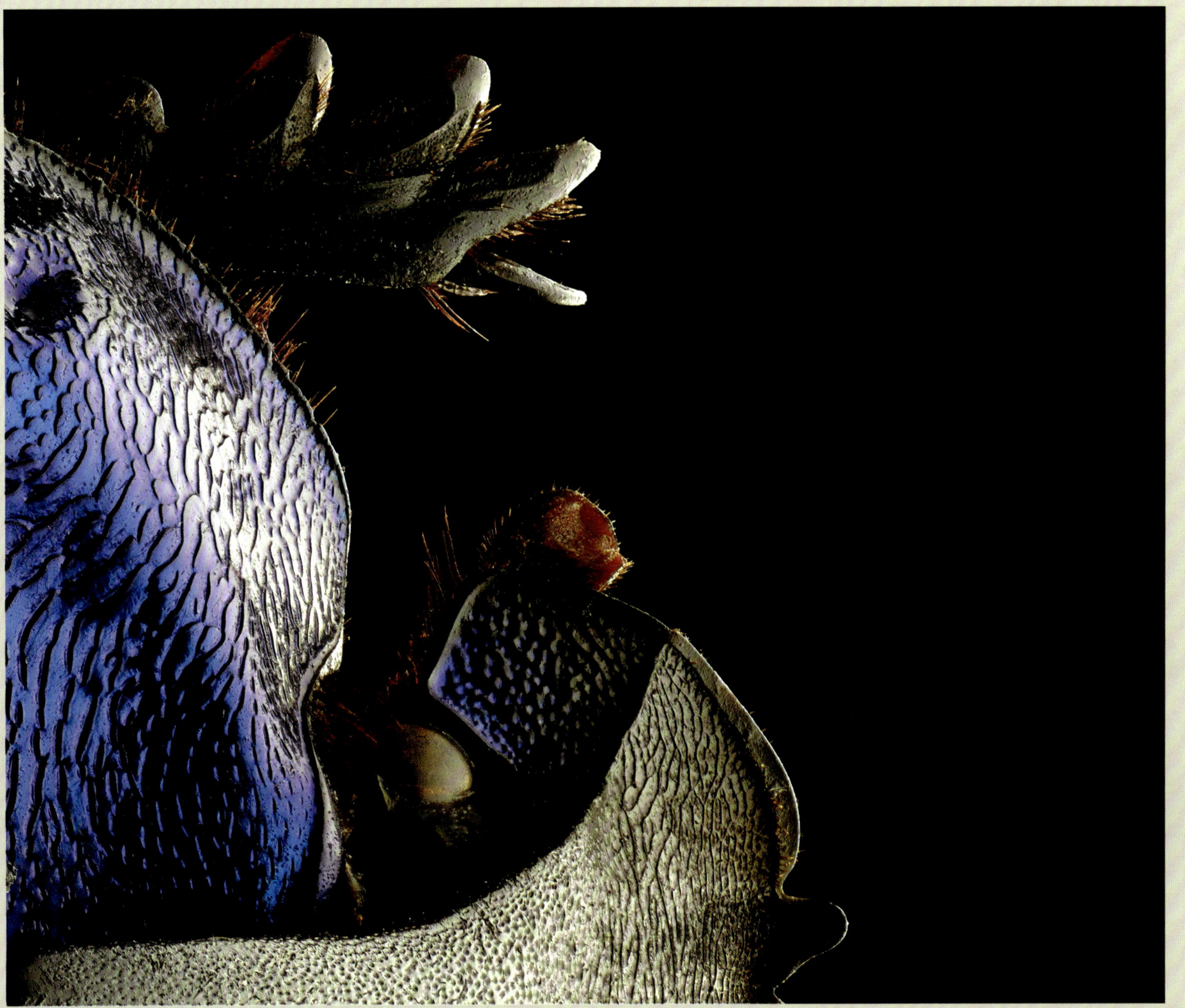

IRIDESCENT BARK MANTIS

Metallyticus splendidus (Mantodea, Metallyticidae)

The spectacular metallic colors of this species are very unusual for a mantis; also atypical is the wide front femur armed with a large, dagger-like spine. The position of *Metallyticus* in the mantis tree of life is not yet clear, but its short body and other features resemble some primitive fossil species. Recent evidence from DNA sequences has confirmed that mantises are close relatives of cockroaches.

MALAYSIA

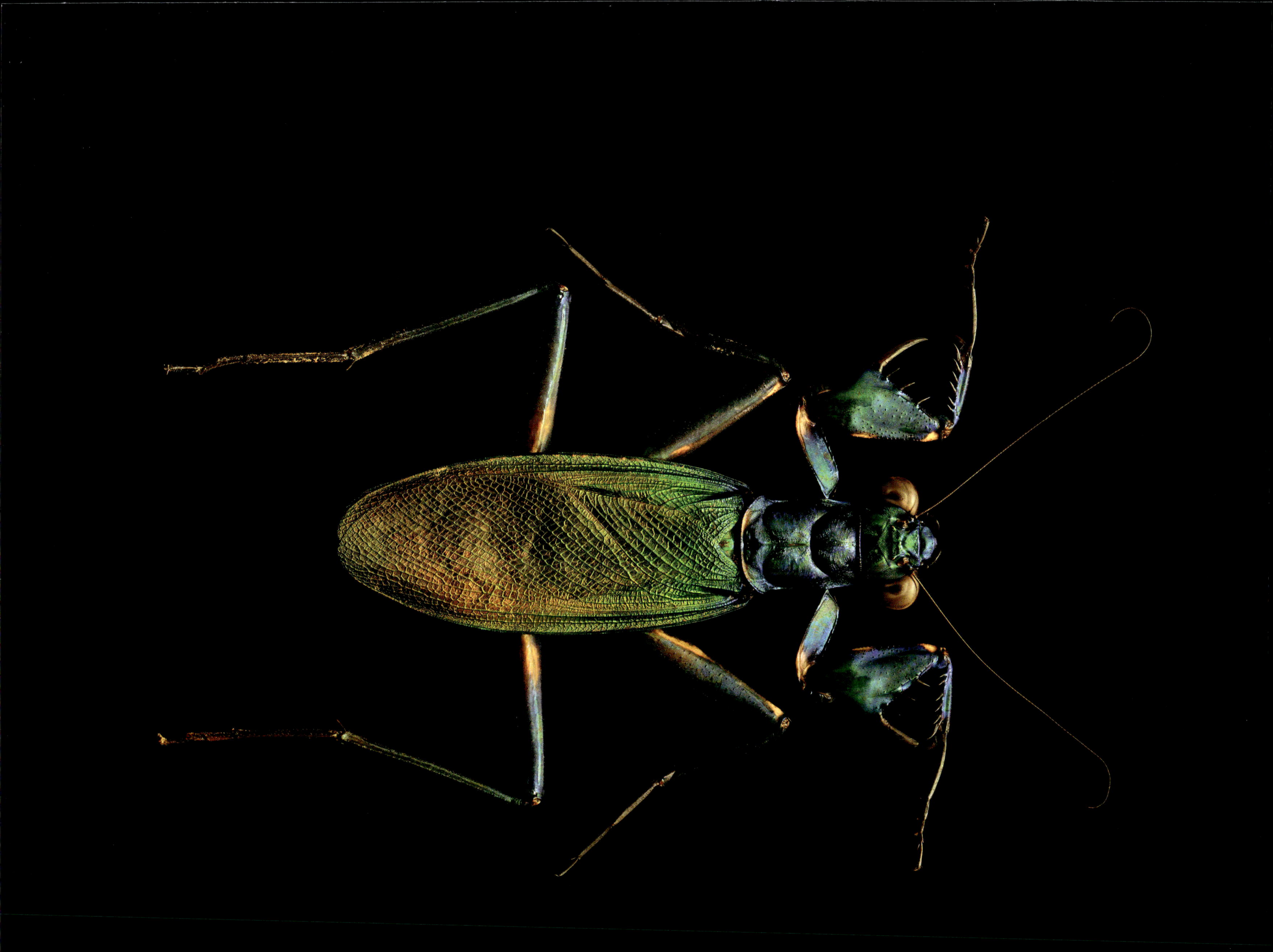

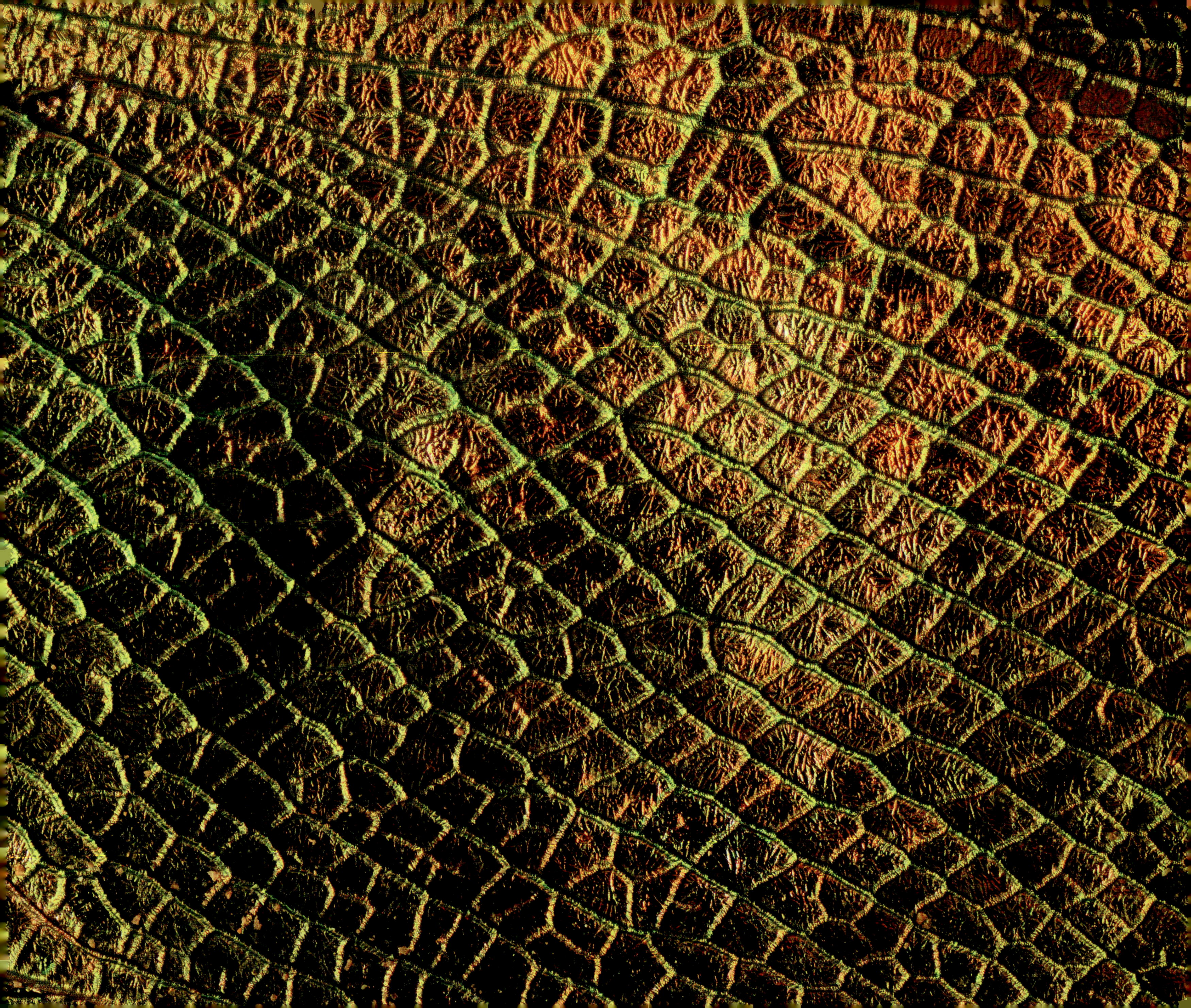

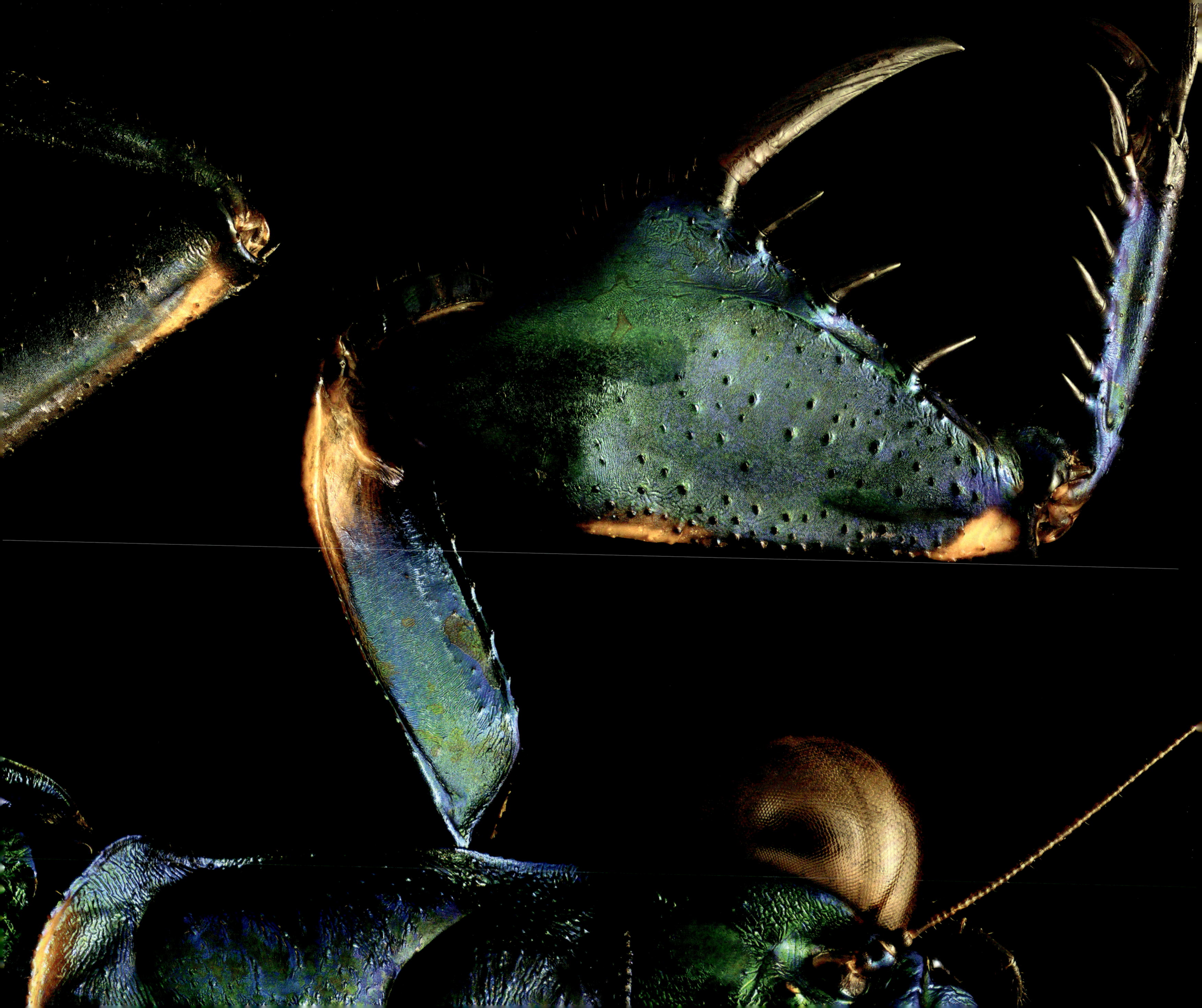

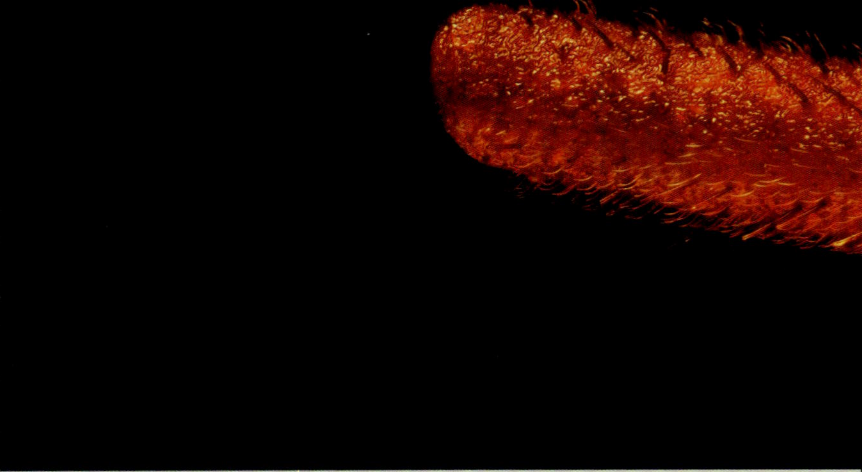

SHIELD BUG

SPECIES OF THE SUPERFAMILY *Pentatomoidea* (Hemiptera)

This shield bug was collected by Charles Darwin in 1836 during his time aboard the HMS *Beagle*. Considering its age and how far it has traveled, both in distance and time, it is in remarkable condition. This is one of many specimens in the Darwin collection of the Oxford University Museum. The founder of the insect collection at Oxford, the Reverend F.W. Hope, was a friend and mentor to Darwin, corresponding regularly and traveling with him on several occassions to collect beetles.

AUSTRALIA

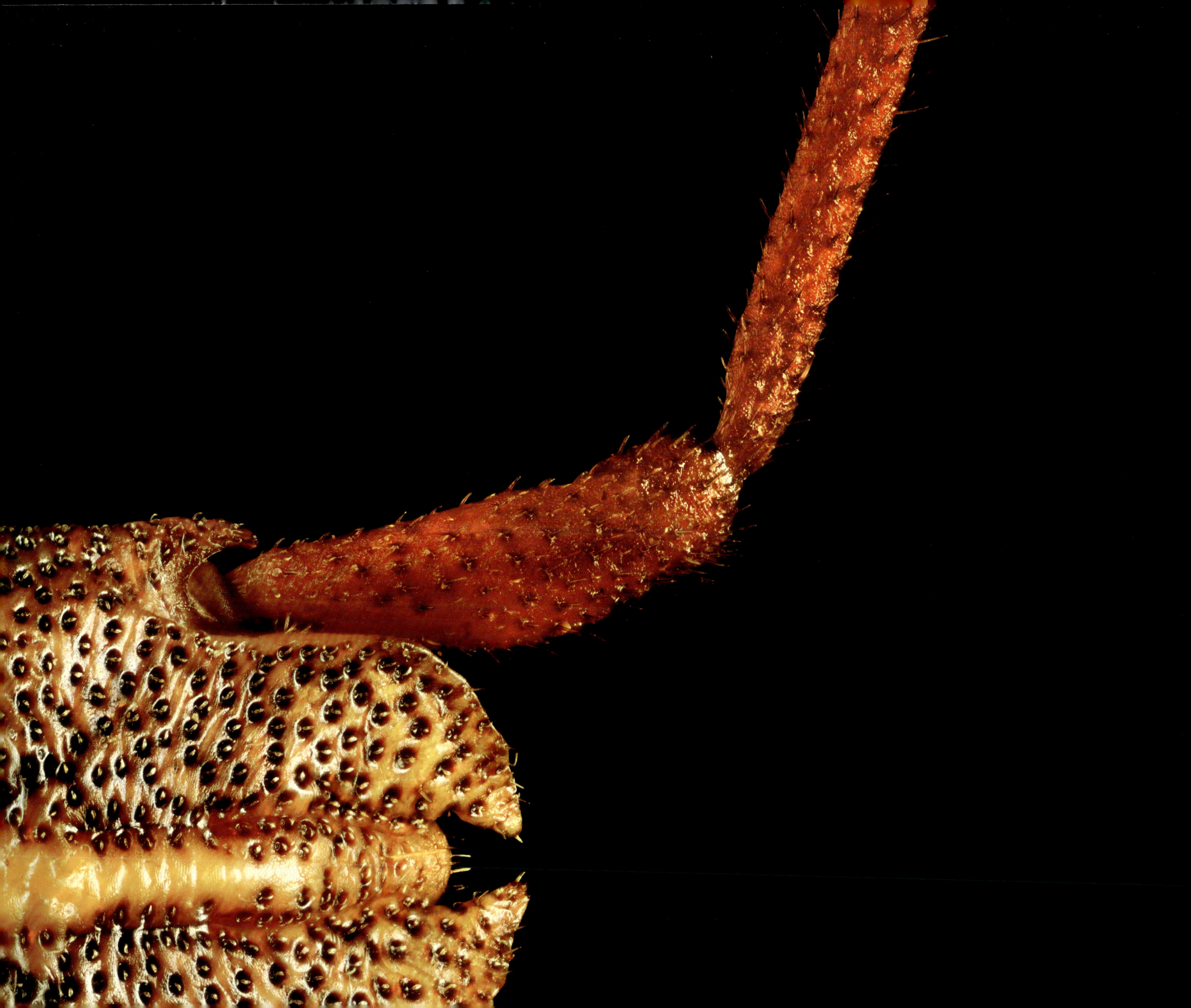

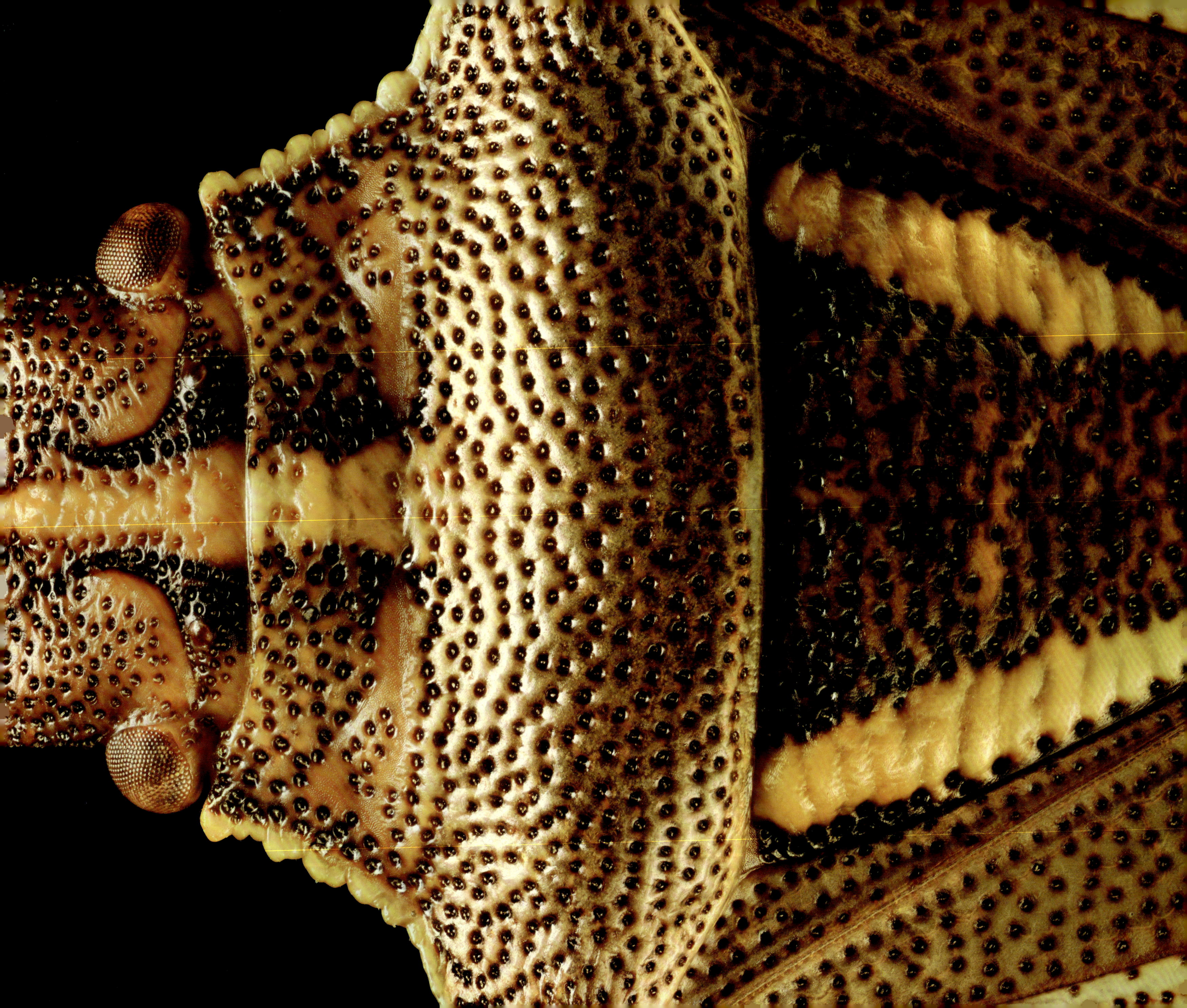

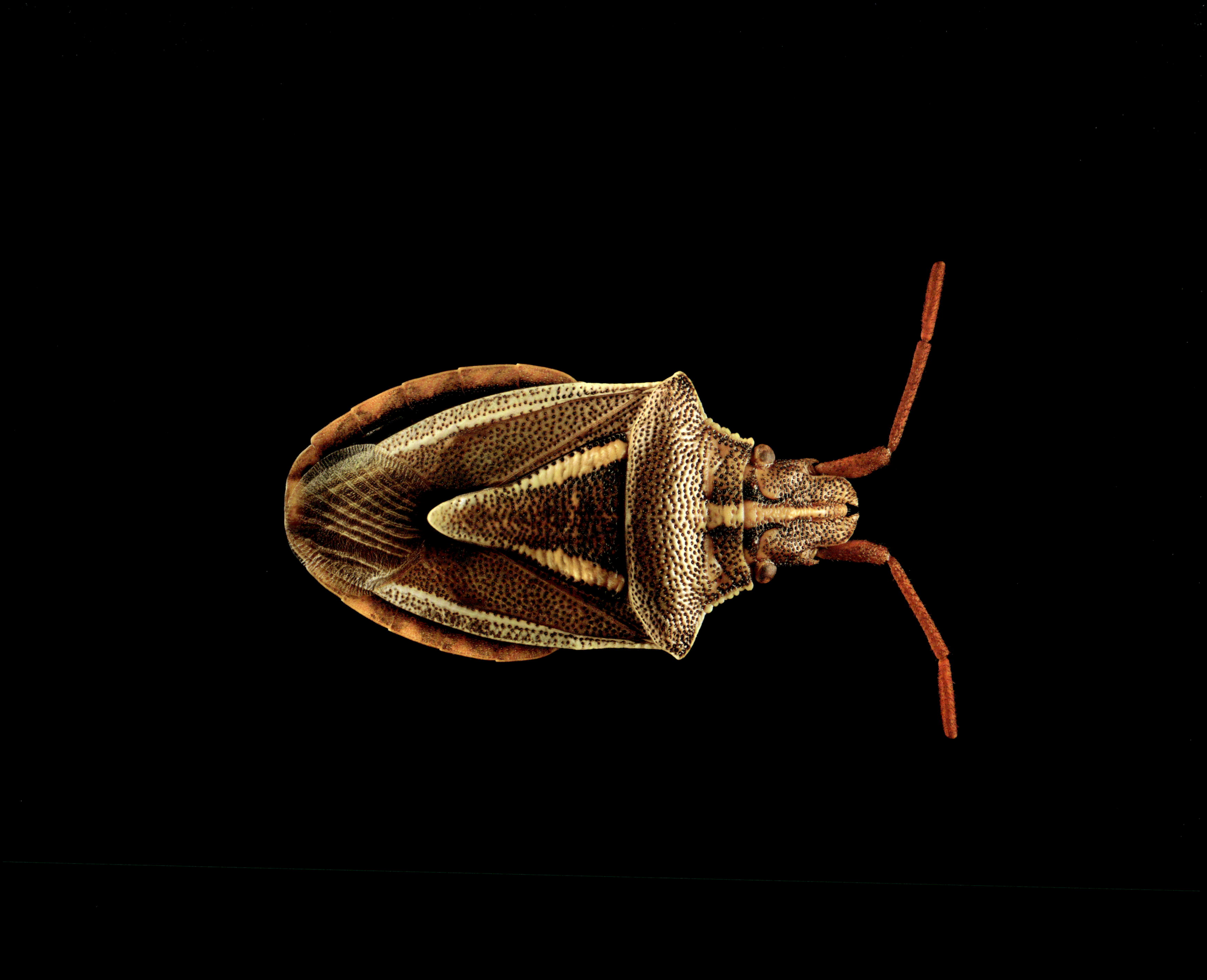

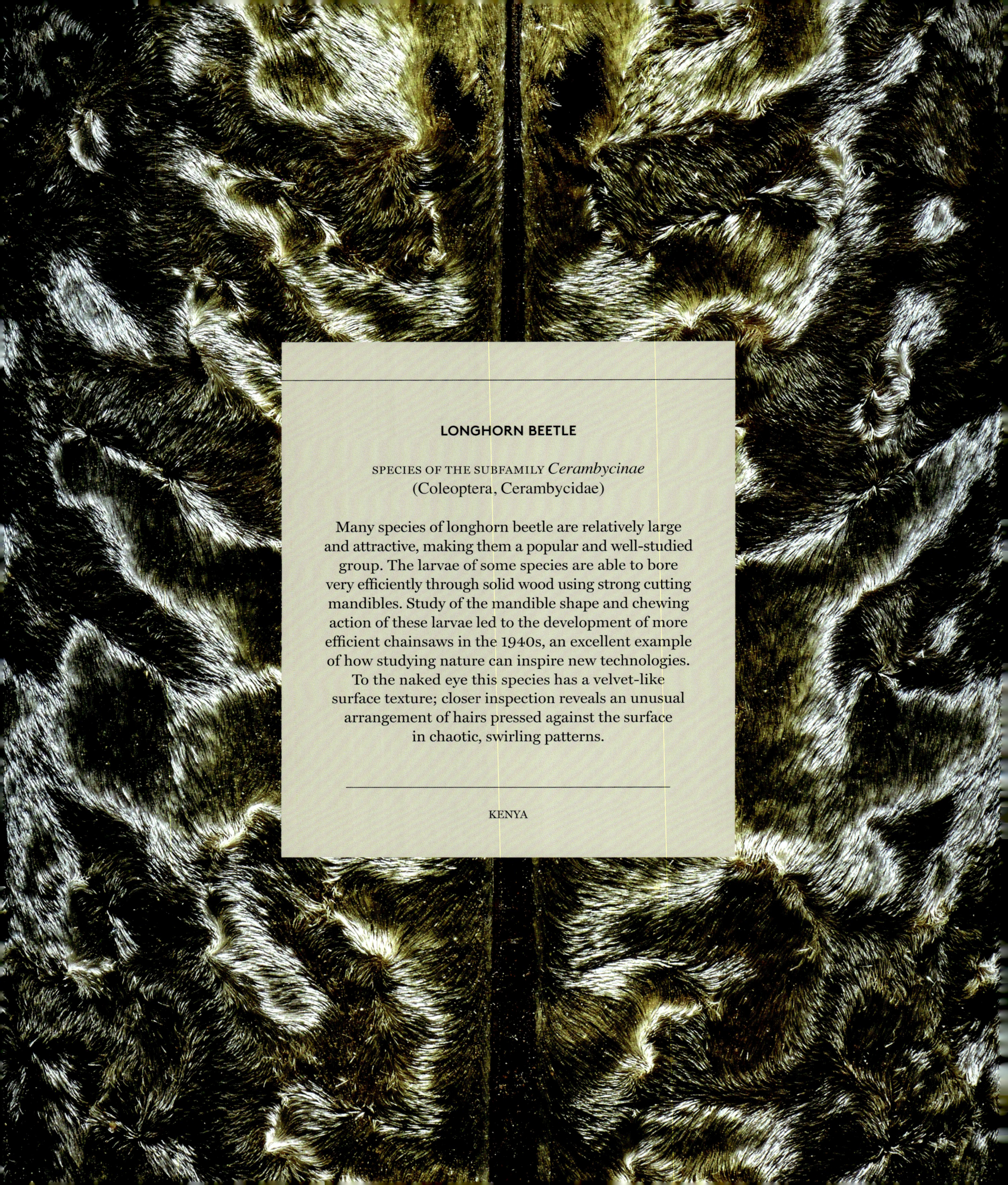

LONGHORN BEETLE

SPECIES OF THE SUBFAMILY *Cerambycinae*
(Coleoptera, Cerambycidae)

Many species of longhorn beetle are relatively large and attractive, making them a popular and well-studied group. The larvae of some species are able to bore very efficiently through solid wood using strong cutting mandibles. Study of the mandible shape and chewing action of these larvae led to the development of more efficient chainsaws in the 1940s, an excellent example of how studying nature can inspire new technologies.

To the naked eye this species has a velvet-like surface texture; closer inspection reveals an unusual arrangement of hairs pressed against the surface in chaotic, swirling patterns.

KENYA

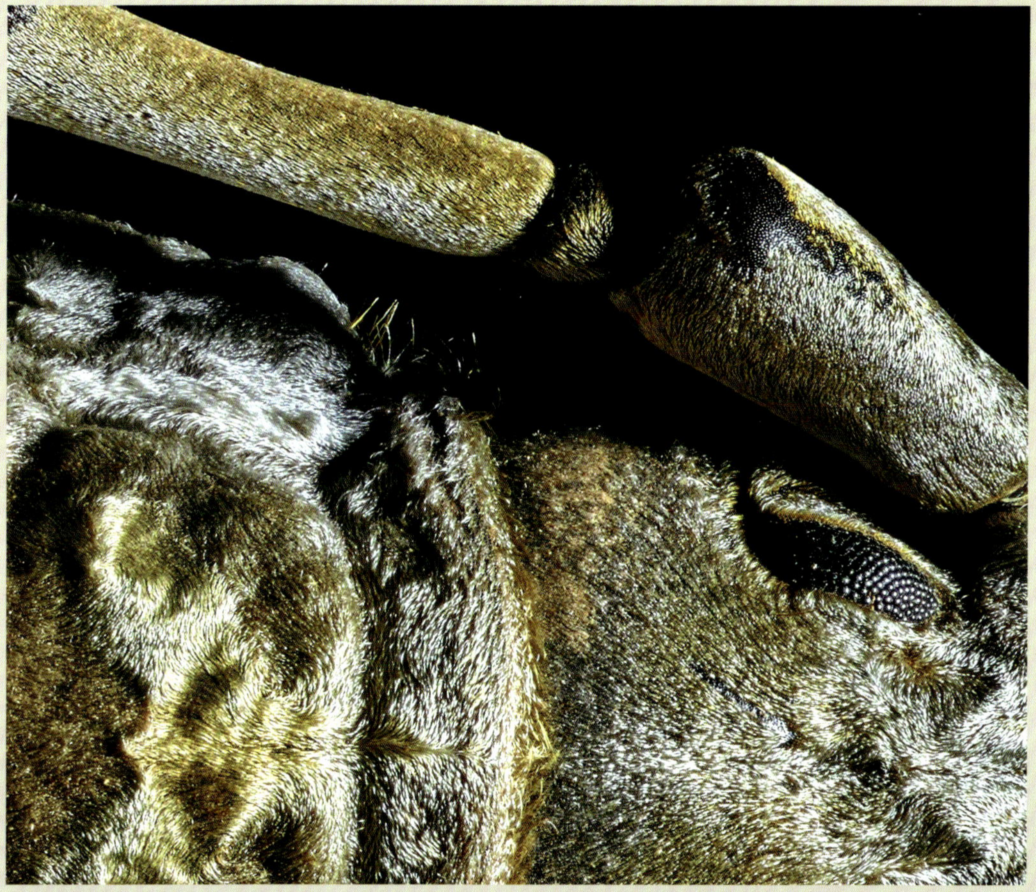
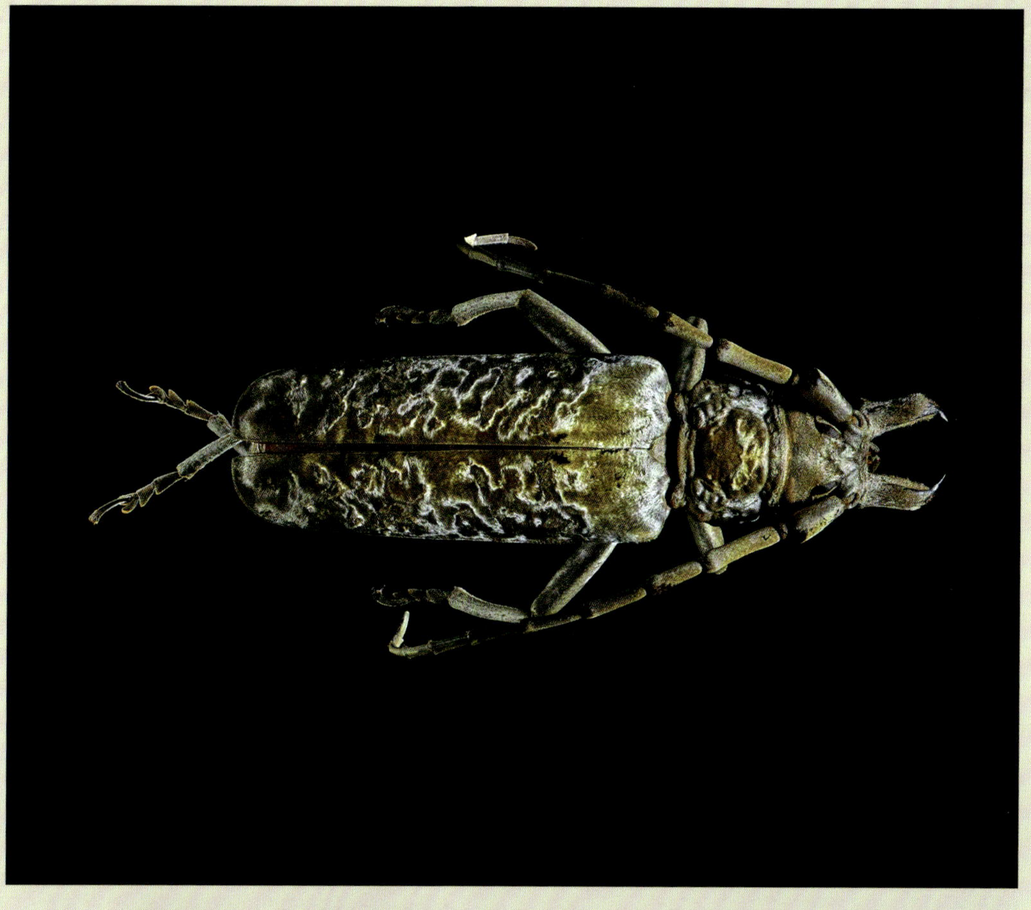
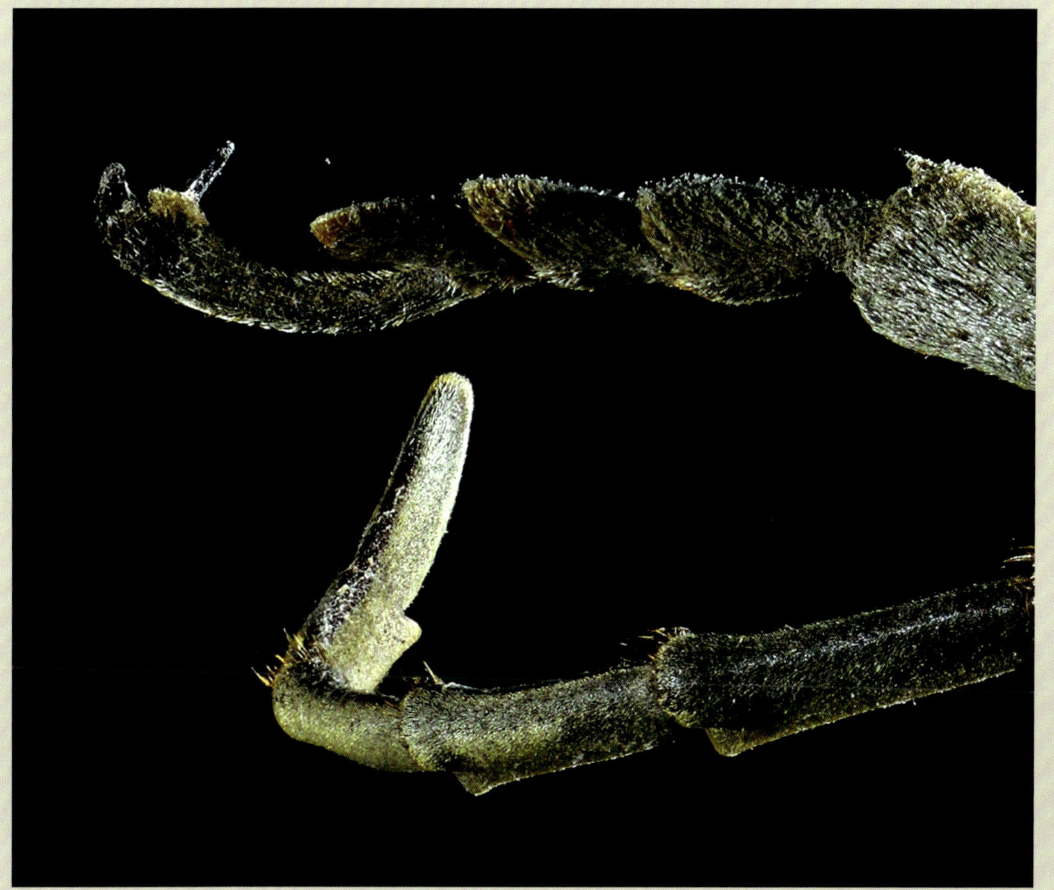
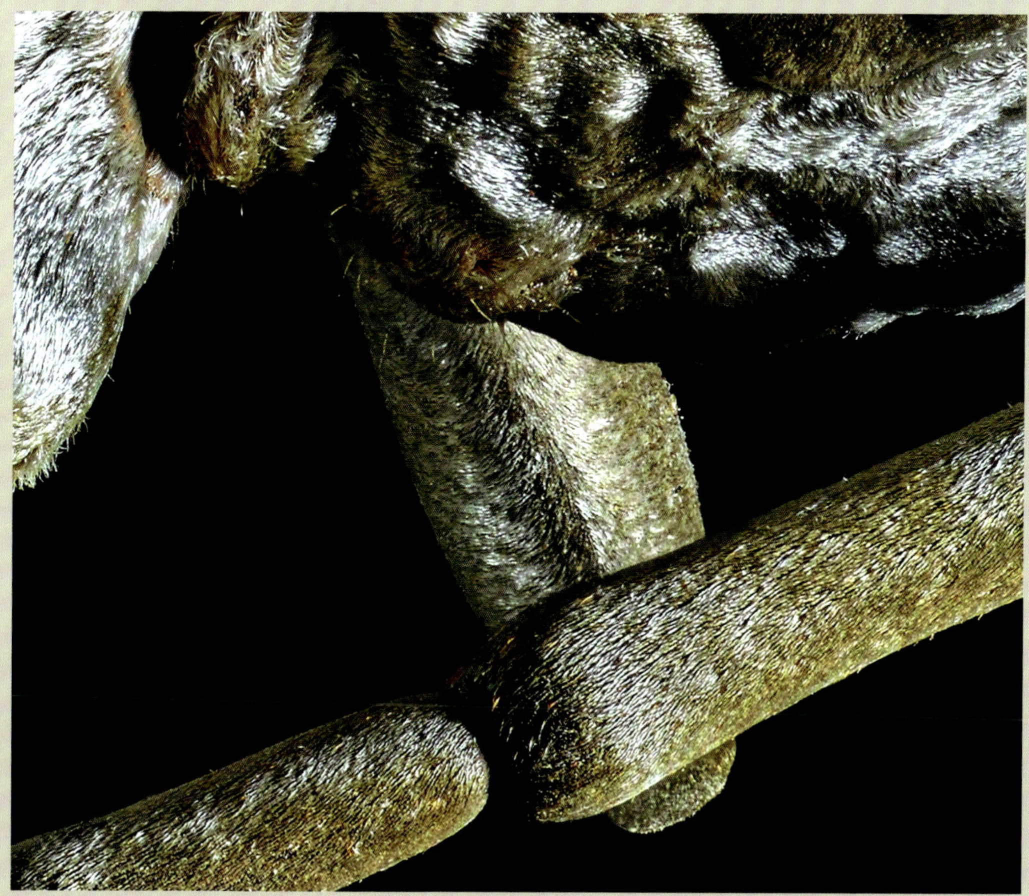

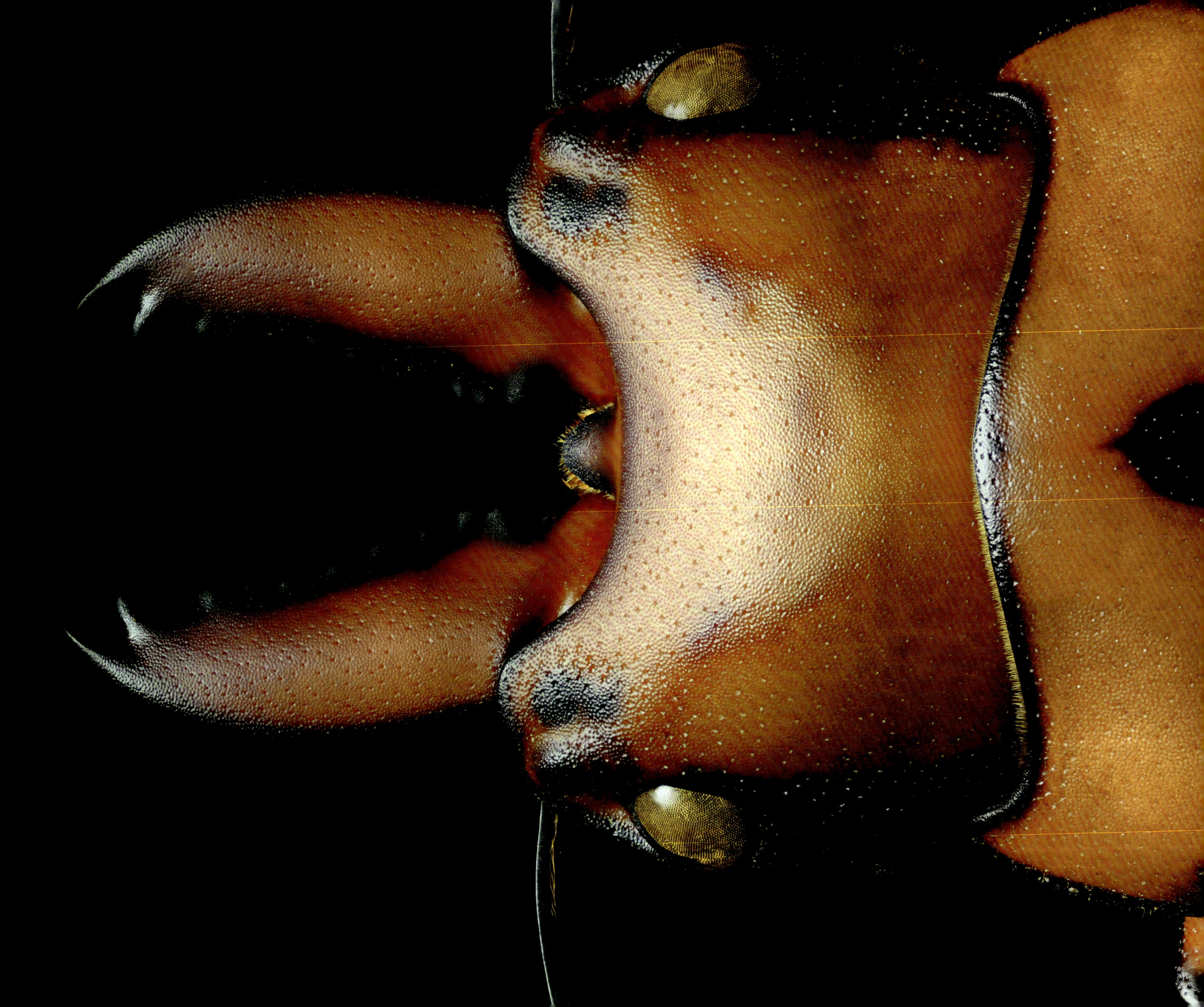

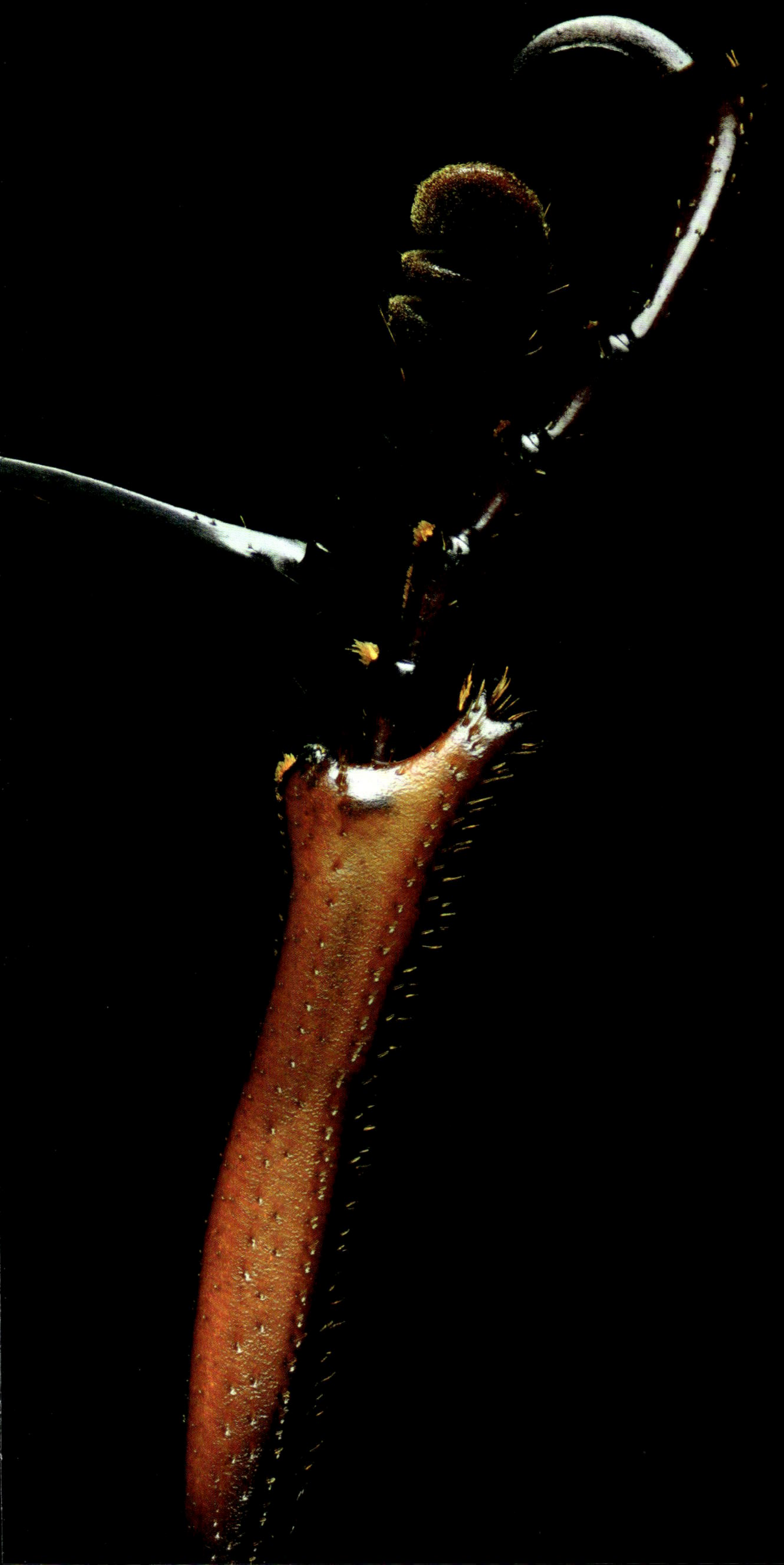

STAG BEETLE

SPECIES OF THE FAMILY Lucanidae (Coleoptera)

Stag beetles are some of the most well known and easily recognized beetles. In competing for mates, males use their enlarged mandibles for grappling rivals. Adult stag beetles are relatively short-lived, spending most of their lives in the larval stage consuming rotting wood. These and other wood-boring beetles play an important role in forest ecosystems, physically breaking up dead wood and accelerating decomposition by bacteria and fungi.

SULAWESI, INDONESIA

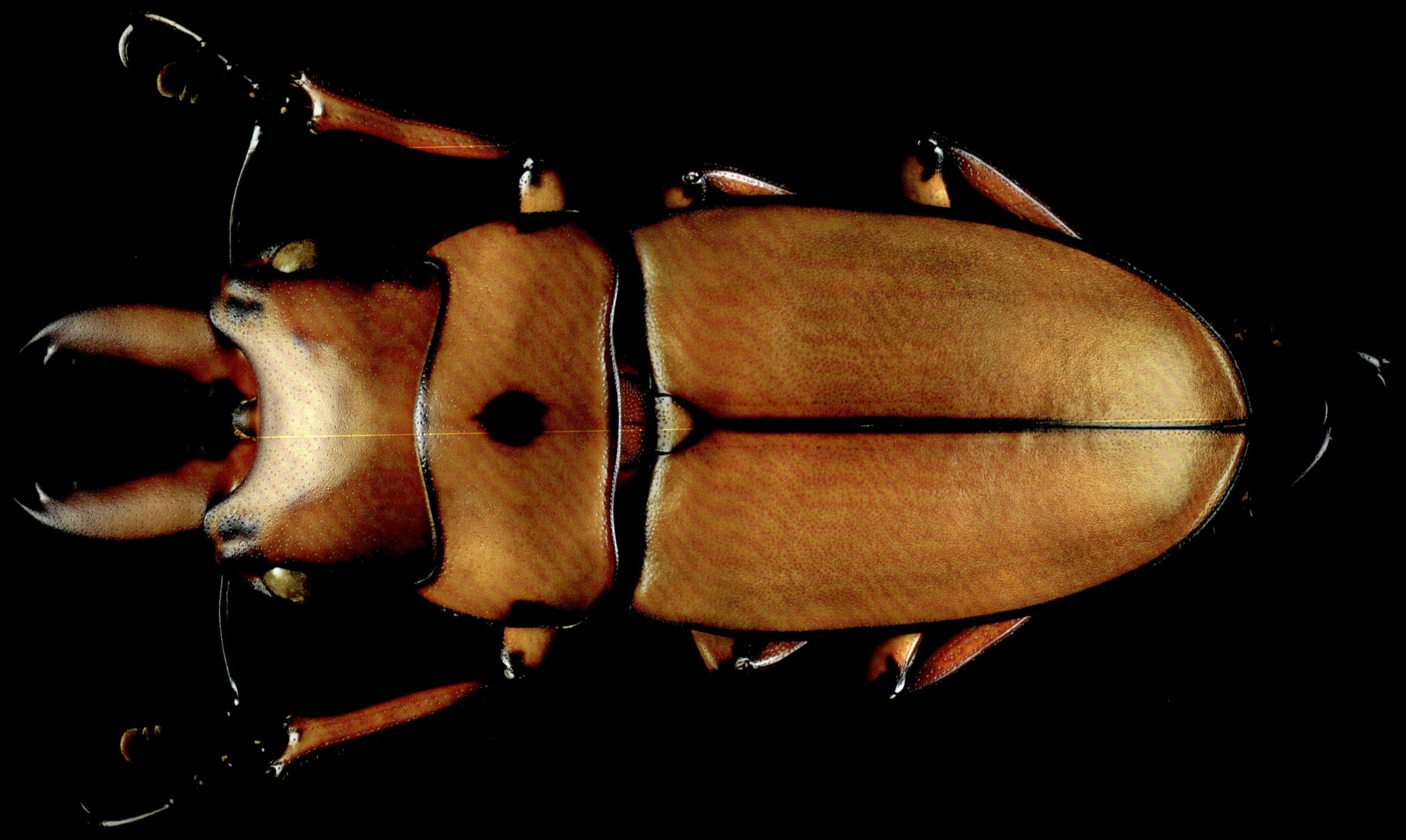

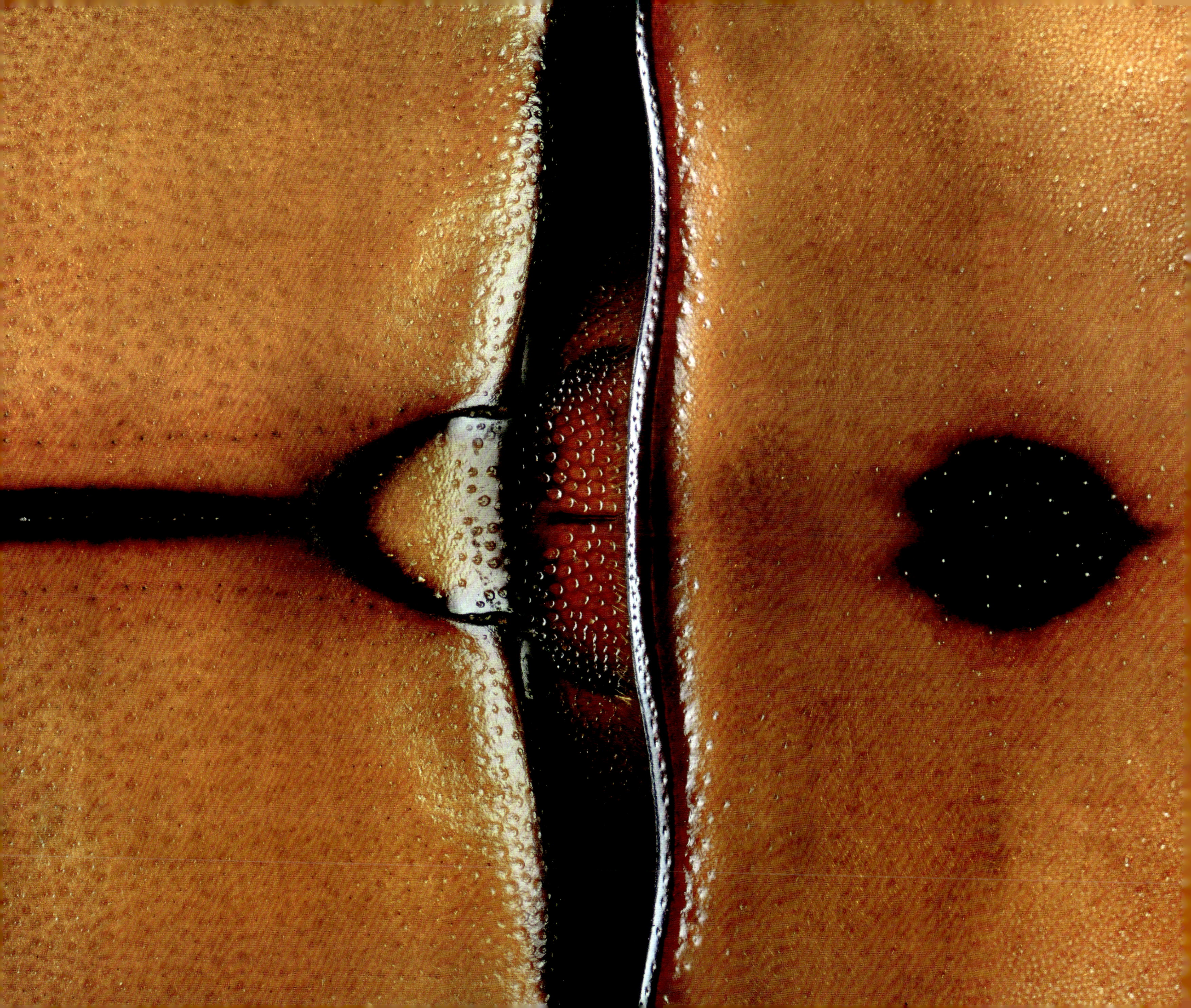

SHORT-NOSED WEEVIL

SPECIES OF THE SUBFAMILY Entiminae
(Coleoptera, Curculionidae)

Weevils are one of the mega-diverse families of insects with over sixty thousand described species. Undoubtedly there are many more awaiting discovery, particulary from tropical forests. This species of short-nosed weevil is endowed with a complex covering of hairs and scales. Backlighting further highlights a fringe of specialized yellow hairs at the end of the legs, which probably help the insect grip smooth surfaces.

BRAZIL

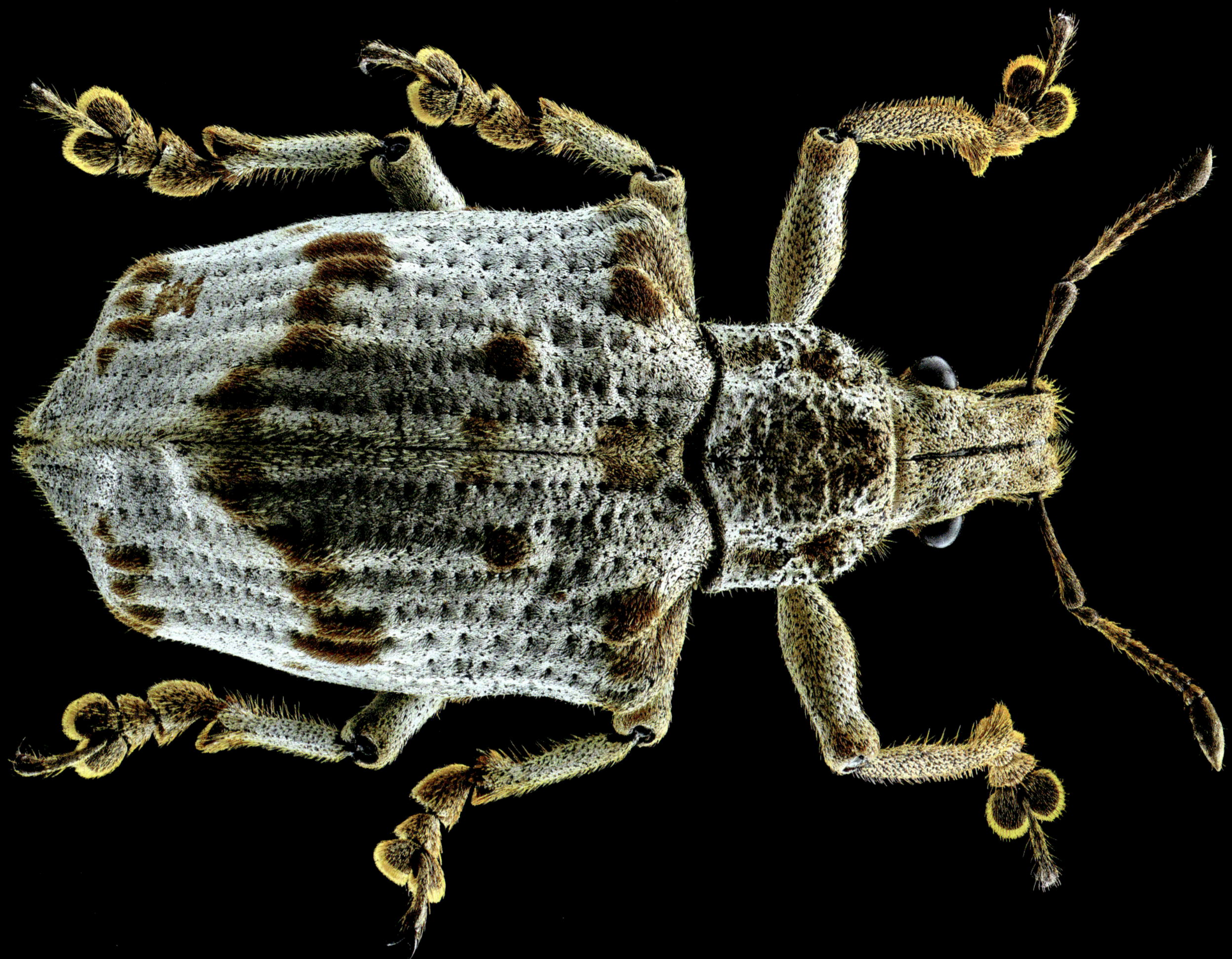

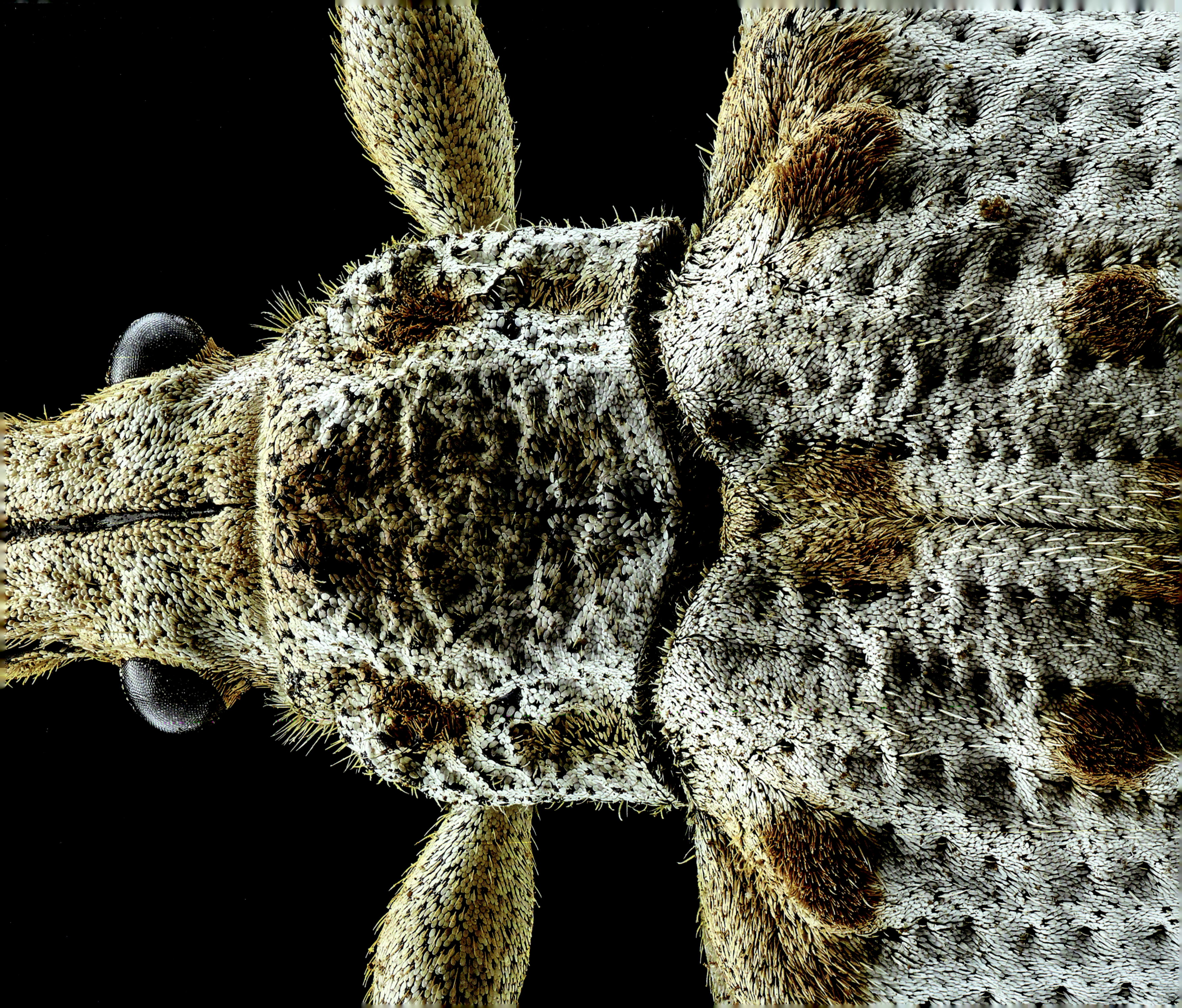

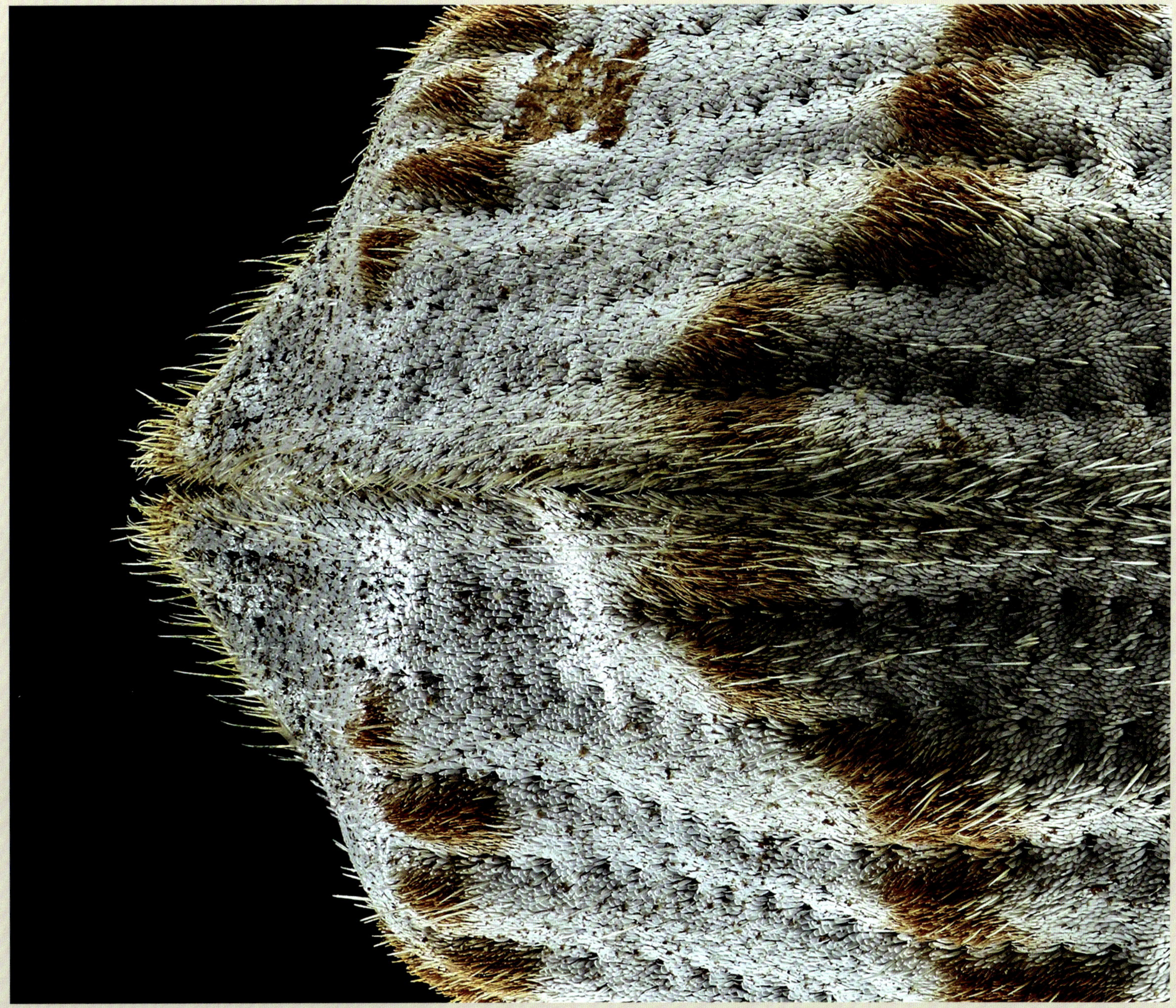

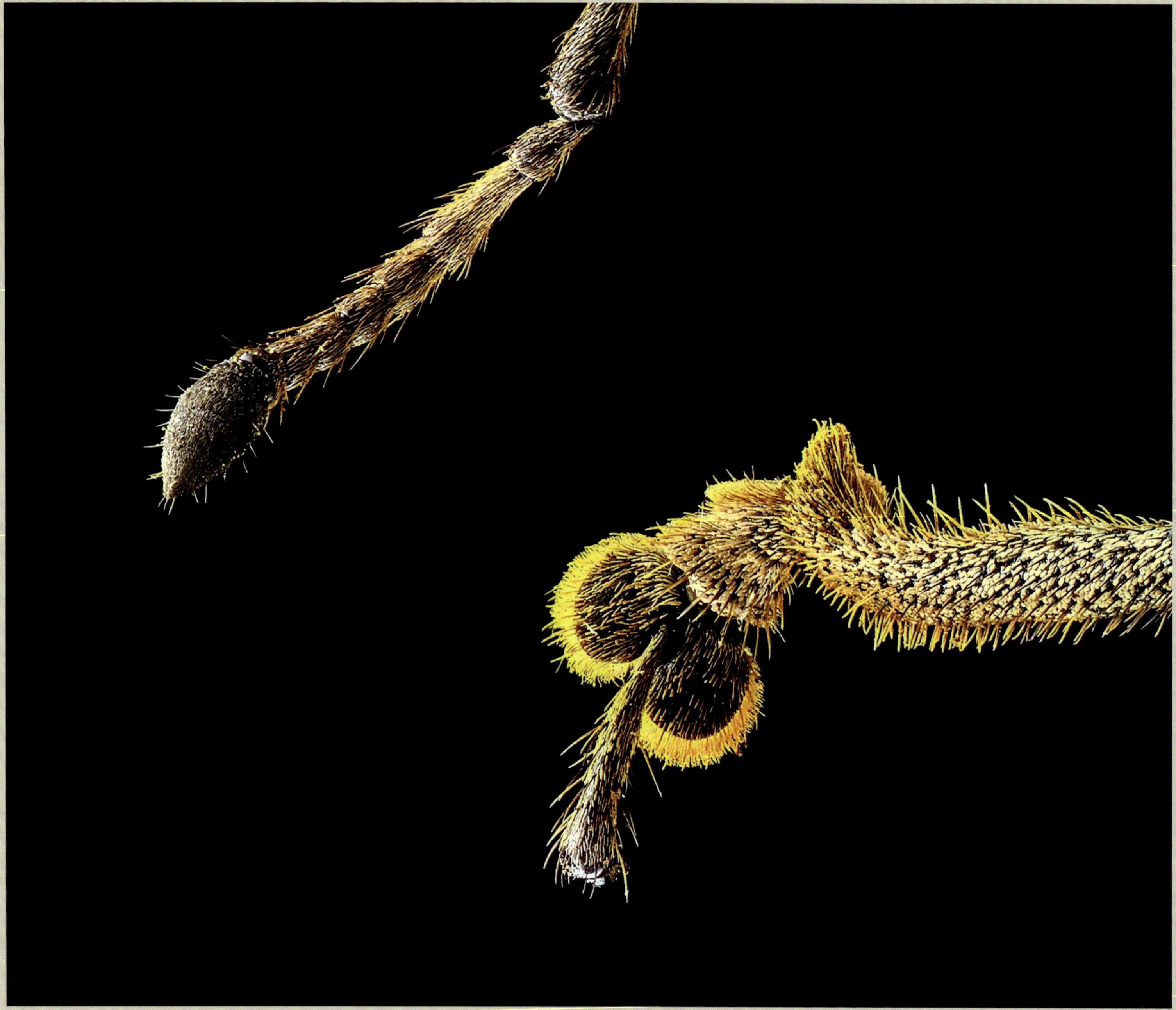

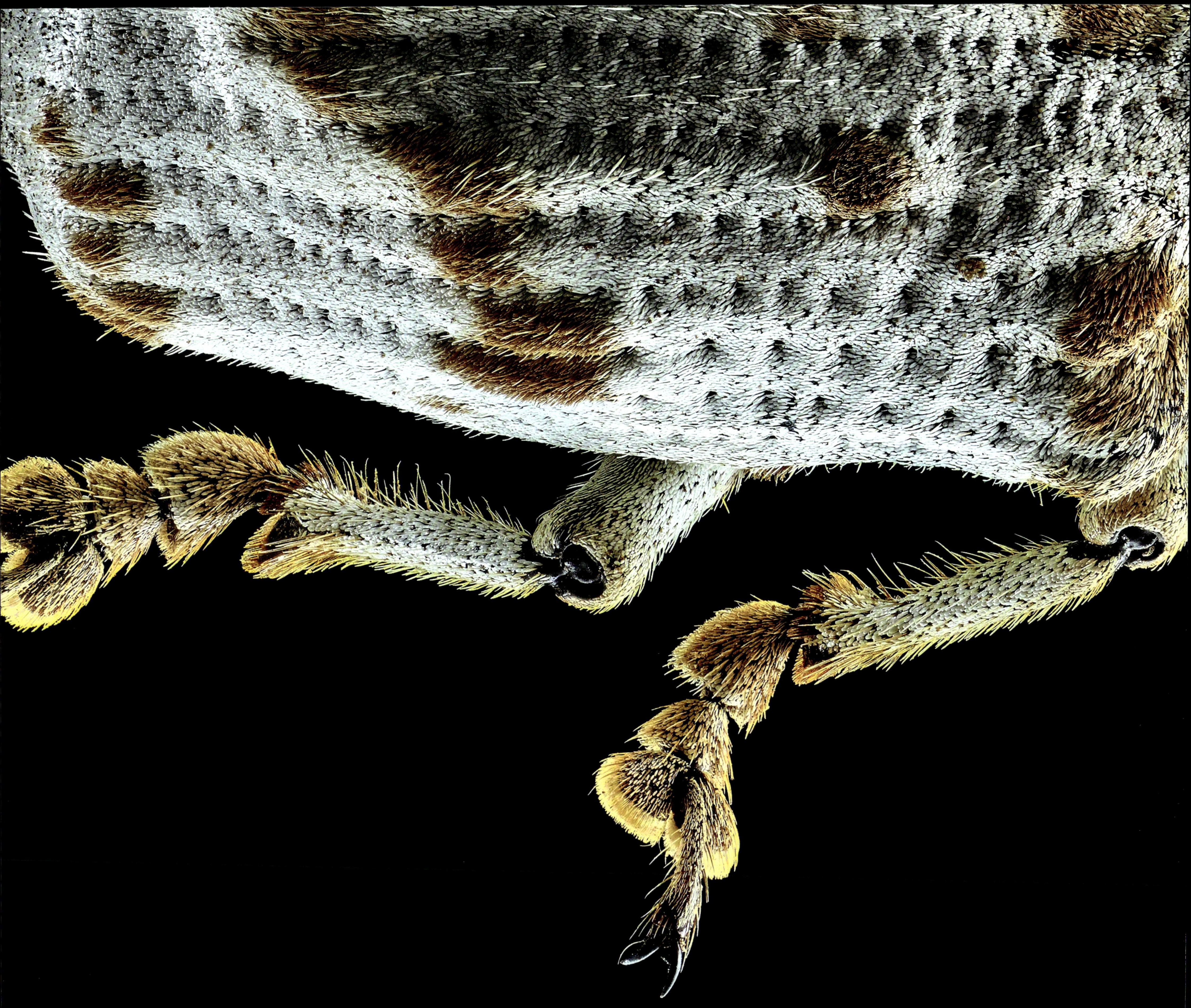

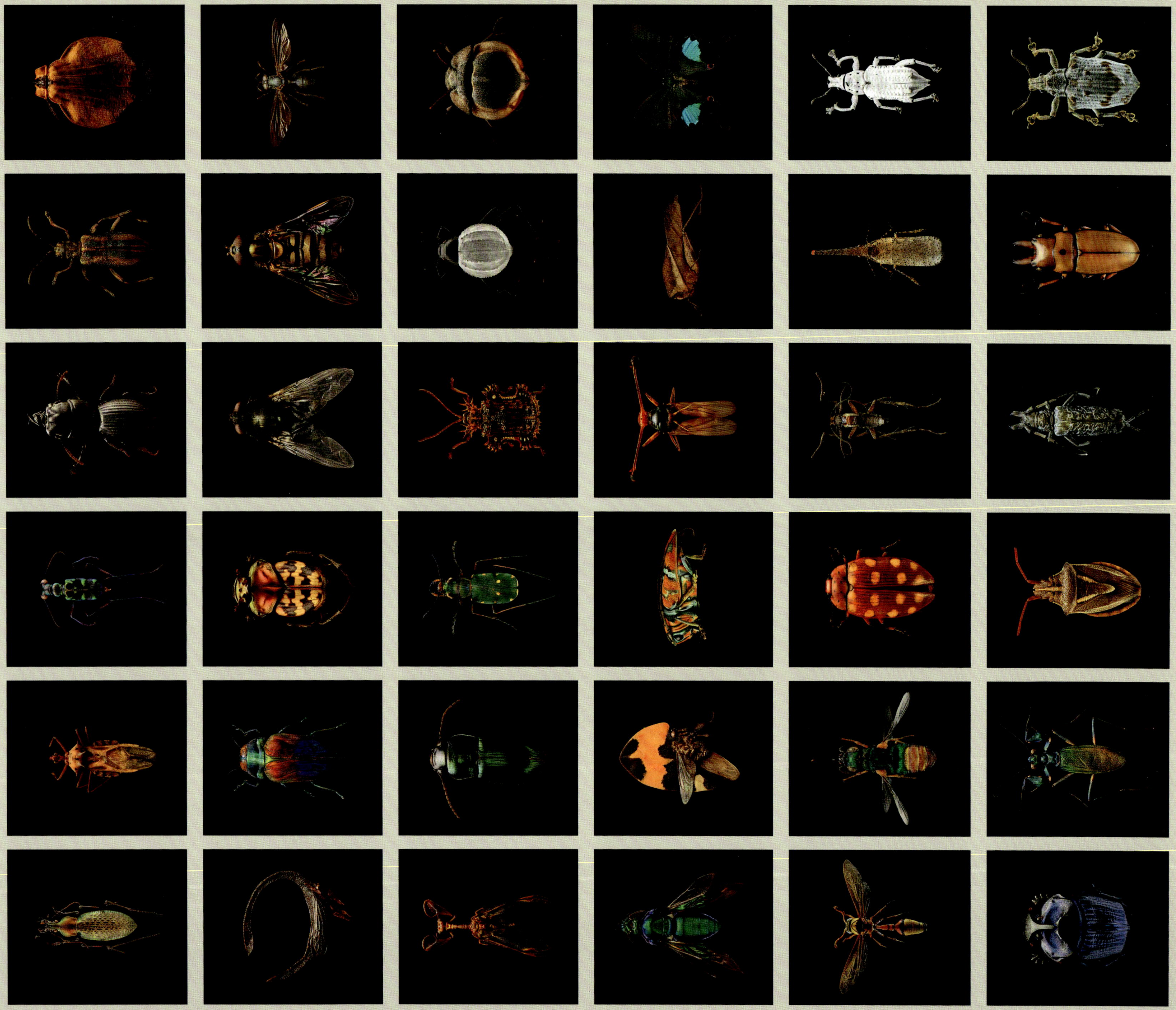

photographic process

I photograph the insect in approximately thirty different sections, depending on the size of the specimen. I light each section differently with strobe lights to bring out the micro-sculptural beauty of that particular section of the insect. For example, I will light and shoot just one antenna, then I will move on to the eye and the lighting set up will change entirely to suit the texture and contours of that part of the body. This process continues until I have covered the whole surface area of the insect.

Due to the inherent shallow depth of field that microscope lenses provide, each individual photograph only contains a tiny slither of focus. To enable me to capture all the information I need to create a fully focused image, the camera is mounted onto an electronic rail that I program to move forward ten microns between each shot. To give you an idea of how far that is, the average human hair is around seventy-five microns wide. The camera will then slowly move forward from the front of the insect to the back creating a folder of images that each have a thin plane of focus. Through various photo-stacking processes I flatten eight thousand images down to create a single picture that has complete focus throughout the full depth of the insect.

I repeat this process over the entire body of the insect and once I have thirty fully focused sections I bring them together in Photoshop to create the final image. From start to finish, a final photograph will take around three weeks to shoot, process, and retouch.

Levon Biss, FRPS

Editor: Garrett McGrath
Designer: John Gall
Production Manager: Anet Sirna-Bruder

Library of Congress Control Number: 2017931140

ISBN: 978-1-4197-2695-8

Printed and bound in China
10 9

ABRAMS
The Art of Books

195 Broadway
New York, NY 10007
abramsbooks.com

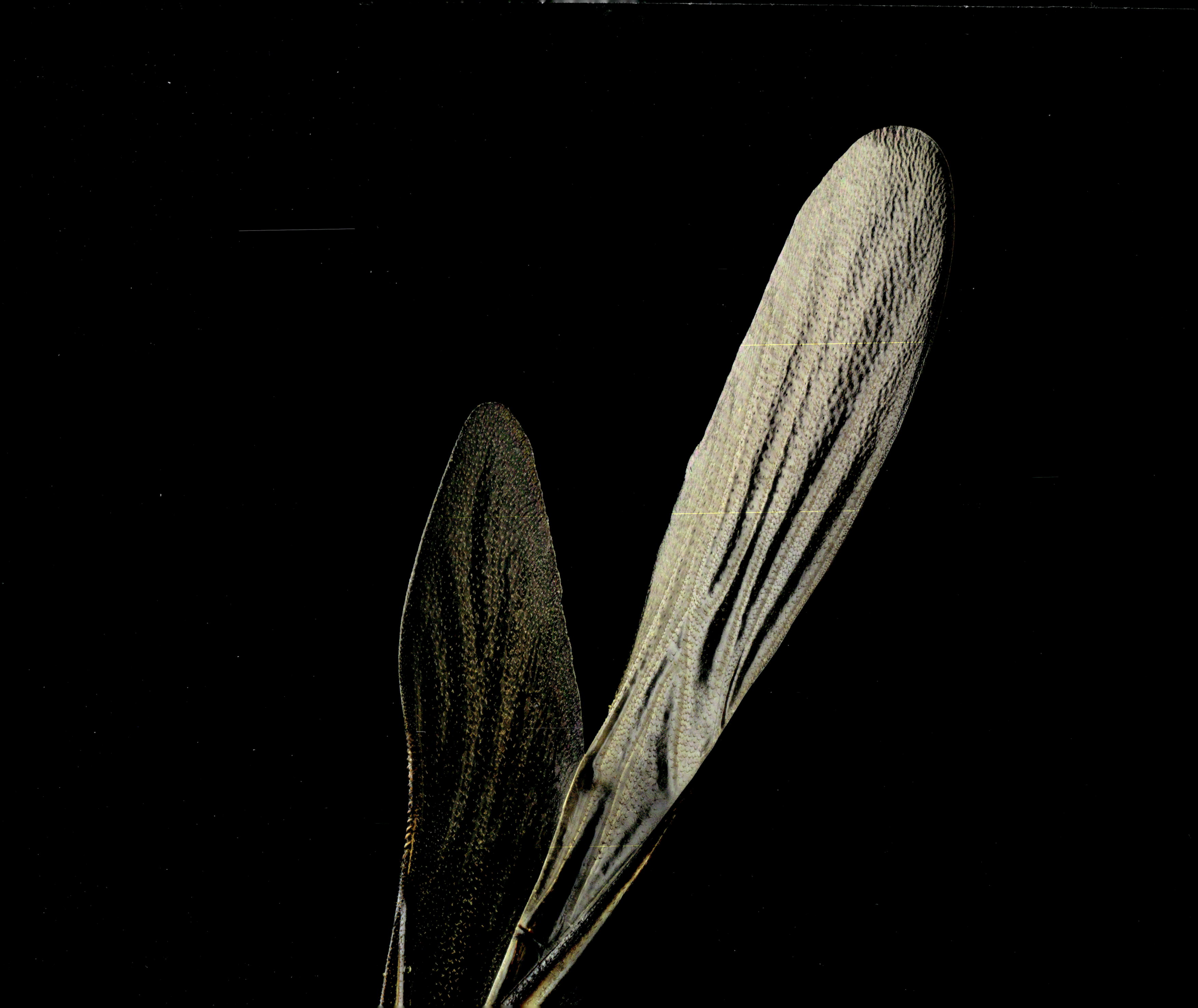